3rd International Scientific and Technical Conference on Scientific and Technical Progress in Ferrous Metallurgy (SATPIFM 2017)

IOP Conference Series: Materials Science and Engineering
Volume 287

Cherepovets, Russia
19 – 20 October 2017

ISBN: 978-1-5108-5502-1
ISSN: 1757-8981

Printed from e-media with permission by:

Curran Associates, Inc.
57 Morehouse Lane
Red Hook, NY 12571

Some format issues inherent in the e-media version may also appear in this print version.

Printed by Curran Associates, Inc. (2018)

For permission requests, please contact the Institute of Physics
at the address below.

Institute of Physics
Dirac House, Temple Back
Bristol BS1 6BE UK

Phone: 44 1 17 929 7481
Fax: 44 1 17 920 0979

techtracking@iop.org

Additional copies of this publication are available from:

Curran Associates, Inc.
57 Morehouse Lane
Red Hook, NY 12571 USA
Phone: 845-758-0400
Fax: 845-758-2633
Email: curran@proceedings.com
Web: www.proceedings.com

Table of contents

Volume 287

3rd International Scientific and Technical Conference on Scientific and Technical Progress in Ferrous Metallurgy - SATPIFM 2017

19–20 October 2017, Cherepovets, Russia

Accepted papers received: 14 December 2017
Published online: 2 January 2018

Preface

3rd International Scientific and Technical Conference on Scientific and Technical Progress in Ferrous Metallurgy - SATPIFM 2017

Peer review statement

Papers

Development of the Mathematical Model for Ingot Quality Forecasting with Consideration of Thermal and Physical Characteristics of Mould Powder

K N Anisimov, A M Loginov, M P Gusev, S V Zarubin, S V Nikonov and A V Krasnov.....1

Visualization of Process of Wheel Steel High Ingots Simulation

V I Bondarenko, V V Bodryaga, F V Nedopekin and V V Belousov.....10

Ways to Improve the Quality of Die Steel 5KhNM

S V Efimov, O Yu Malykhina, A G Pavlova, V G Milyuts, V V Tsukanov and V V Vikharev.....18

New Engineering Solutions in Creation of Mini-BOF for Metallic Waste Recycling

S P Eronko, S M Gorbatyuk, E V Oshovskaya and B I Starodubtsev.....25

Analyzing Operating Behavior of Hot Mill Table Roll Drives using Statistical Methods for Current Values

P Z Kabakov, A V Kozhevnikov and I S Ilatovsky.....30

Investigation of the Change in the Magnetic Properties of Die Steels in the Process of Manufacture and Operation of Die Tools

I I Karpukhin, A I Selishchev, A V Kornilova and I M Idarmachev.....35

The Study of Influence of Technology Factors and Equipment Design Parameters on Energy Consumption taking into Account Stochasticity of the Cold Rolling Process

A V Kozhevnikov, N L Bolobanova and I A Kozhevnikova.....42

Competences of Engineers in the Iron and Steel Industry

A V Kozhevnikov.....47

Intelligent Support System of Steel Technical Preparation in an Arc Furnace: Functional Scheme of Interactive Builder of the Multi Objective Optimization Problem

O S Logunova and N S Sibileva.....52

Study of the Impact of Heat Treatment Modes on Formation of Microstructure and a Given Set of Mechanical Properties of High-Strength Flat Products with Guaranteed Hardness (400 to 450 HB) from Low-Alloyed Steel

M Yu Matrosov, P G Martynov, T V Goroshko, M I Zvereva, A V Mitrofanov and K Yu Barabash.....56

Experience of Application of Liquid Lubricating Materials during Wide Strip Hot Rolling

S I Platov, R R Dema, M V Kharchenko and R N Amirov.....62

Economical and Energy Efficiency of Iron and Steel Industry Reindustrialisation in Russia Based on Implementation of Breakthrough Energy-Saving Technologies

L N Shevelev.....67

Scientific and Technological Principles of Development of New Cold-Resistant Arc-Steels (Steels for Arctic Applications)

O V Sych, E I Khlusova and E A Yashin.....70

Numerical Simulation of Roller Levelling using SIMULIA Abaqus

K A Trusov, P A Mishnev, O V Kopaev and D V Nushtaev.....76

Optimization of Thermal and Deformation Effect during Plastic Deformation and Thermal Treatment of Hot Rolled Heavy Wall Pipes Made of Medium Carbon Martensitic Steel

A I Ziza and V V Tsukanov.....80

Foreword

The 3rd International Scientific and Technical Conference on Scientific and Technical Progress in Ferrous Metallurgy (SATPIFM 2017) was held from October 19th to 20th, 2017 in Cherepovets, Vologda Region, Russia. The organizers of the Conference are PAO Severstal and Cherepovets State University.

SATPIFM is the regular international conference series in Russia. It has been held in Cherepovets biennially since 2013.

SATPIFM is designed to bring together scientists, engineers, researchers in the field of ferrous metallurgy, as well as representatives of metallurgical companies for sharing ideas and fostering research network and worldwide collaboration.

Over 120 representatives from 7 countries (Russia, Ukraine, Sweden, Italy, Czech Republic, Spain and France), 11 universities (Cherepovets State University, Higher School of Economics, Yaroslavl State University, Moscow Institute of Steel and Alloys, Bauman Moscow State Technical University, Nosov Magnitogorsk State Technical University, Lomonosov Moscow State University, South Ural State University, Stankin Moscow State Technological University, Donetsk National Technical University, Moscow Aviation Institute), 6 manufacturing companies (Severstal, Magnitogorsk Iron and Steel Works, Vyksa Steel Works, Novokramatorsk Machine Building Plant, Chepetsky Mechanical Plant, Novolipetsk Steel Plant) and 20 research institutions, design centers, research and production associations (Bardin Central Research Institute of Ferrous Metallurgy, National Research Center “Kurchatovsky Institute”, Central Research Institute of Structural Materials "Prometey", CHELYABGIPROMEZ (Chelyabinsk National Design Institute for Metallurgical Plants), Research and Development Center “High Technology Metallurgy”, Institute of Ferrous Metallurgy of the National Academy of Sciences of Ukraine, Royal Institute of Technology (Sweden), FIVES KEODS (France), Research and Development Center for Computer-Aided Engineering Tools, Skolkovo Institute of Science and Technology, IT-Service, Severstal-Infocom and others) took part in the conference.

This collection includes the papers selected from a large number of papers and presentations received by us.

We would like to express our appreciation to the International Organizing Committee and the Local Organizing Committee for good planning of the event, engaging presentations and innovative approach to work. We extend our gratitude to all those involved in organizing the conference for their efficient and hard work, to all the authors who have submitted their works to SATPIFM 2017, and also to the reviewers who provided their valuable comments and suggestions.

Editors
Ph.D. Aleksandr Kozhevnikov
Ph.D. EkaterinaTselikova
Cherepovets State University, Russia

IOP Conf. Series: Materials Science and Engineering **287** (2017) 011002 doi:10.1088/1757-899X/287/1/011002

Peer review statement

All papers published in this volume of *IOP Conference Series: Materials Science and Engineering* have been peer reviewed through processes administered by the proceedings Editors. Reviews were conducted by expert referees to the professional and scientific standards expected of a proceedings journal published by IOP Publishing.

IOP Conf. Series: Materials Science and Engineering **287** (2017) 012001 doi:10.1088/1757-899X/287/1/012001

Development of the Mathematical Model for Ingot Quality Forecasting with Consideration of Thermal and Physical Characteristics of Mould Powder

K N Anisimov[1], A M Loginov[1], M P Gusev[2], S V Zarubin[3], S V Nikonov[4], A V Krasnov[4]

[1] Korad LLC, 23/9 2nd Baumanskaya Street, Moscow, RU-105005, Russia
[2] Skolkovo Institute of Science and Technology, 3 Nobel Street, Moscow, RU-143026, Russia)
[3] Bauman Moscow State Technical University, 5/1 2nd Baumanskaya Street, Moscow, RU-105005, Russia
[4] PAO Severstal, 30 Mira Street, Cherepovets, RU-162608, Russia

E-mail: akn@corad.ru

Abstract. This paper presents the results of physical modelling of the mould powder skull in the gap between an ingot and the mould. Based on the results obtained from this and previous works, the mathematical model of mould powder behaviour in the gap and its influence on formation of surface defects was developed. The results of modelling satisfactorily conform to the industrial data on ingot surface defects.

1. Introduction

The most important functions of the mould powder are control of heat removal from an ingot to the mould and reduction of the ingot drawing force. Earlier [1], the criteria for mould powder performance evaluation obtained through statistical data processing were determined, and currently these criteria are used in the BOF shop of PAO Severstal and demonstrate good results. However, it does not provide information on the physics of the processes occurring in the mould.

Application of particular mould powder may have different impacts on thermal performance of the mould. Owing to shrinkage of molten slag entering the gap between the ingot being formed and the mould wall, an air gap is formed; and this air gap have a significant impact on total thermal resistance in the system. The largest liquid shrinkage is observed on highly basic mixtures ($CaO/SiO_2>1.2$) [2]. The surface roughness depends on the crystal layer thickness [3] and the crystal morphology [4]. Application of highly basic mixtures reduces the heat flux in the mould, and the process itself is commonly referred to as "soft" cooling. Application of "soft" cooling [1, 5, 6] greatly reduces the number of slabs suffering longitudinal cracks in the peritectic range of products. According to data provided in papers [7–9], the thermal resistance of the forming gas gap between the mould wall and the solid layer of the mould powder skull differs much depending on the research method and is determined for a small amount of mould powder samples. Therefore, it is reasonable to conduct studies of thermal and physical properties of mould powder under specific conditions of PAO Severstal.

Another purpose of the mould powder use is to grease the ingot shell and reduce the friction force in the mould. Friction in the mould can be conditionally divided into liquid friction and dry friction. The liquid friction prevails in the upper part [10] of the mould – in the high temperature zone, and the dry friction prevails in the lower part. According to data [11], the maximal friction appears in the lower part of the mould, that is in the dry friction zone. While the liquid friction is determined by viscosity of the molten slag in the gap, the dry friction is determined by interaction between the solid layer of the mould powder and the forming steel shell. The authors of the paper [12] report the results of the tribological study of the "Steel Ball – Mould Powder Solid Layer" system at temperatures up to 900°C, which can be used for evaluation of the dry friction.

Development of physical and mathematical models describing thermal processes in the ingot-slag-mould system provides the possibility to determine parameters affecting the ingot quality, to evaluate the influence of mould powder characteristics on casting parameters, and to model the mould thermal performance in relation to new conditions of casting.

2. Research Methods and Materials

Industrial mould powders of the basic oxygen steel making process at PAO Severstal (Table 1), which properties are examined in detail in paper [4], were used as the study materials.

Table 1. Physical and chemical characteristics of mould powder.

No.	Al_2O_3	K_2O	Na_2O	MgO	F	CaO/SiO_2	*T* heat (°C)	$\eta_{1,300}$ (Pa·s)
1	6.7	1.53	7.7	<1	7.1	0.89	1,130	0.41
2	4.8	0.25	10.5	2	9.7	0.93	1,025	0.18
3	6.3	0.12	4.5	2.6	6.8	1.22	1,175	0.35

Condition of the mould powder skull in the gap was modelled in two phases. During the first phase, the distribution of the heat flux in the mould was reproduced, and in the second phase the physical modelling of determination of thermal and physical characteristics of the mould powder skull was conducted.

According to the paper [13], the heat flux density can be expressed by the equation:

$$q(z) = q0 - (q0 - q_{mean})\frac{1 - e^{-k\frac{z}{\nu}}}{1 - \frac{\nu}{k \cdot L}\left(1 - e^{-k\frac{z}{\nu}}\right)}, \tag{1}$$

where v is casting speed; z is the distance from the meniscus; L is the height of the surface of contact between the ingot and the mould; q_{mean} is the mean density of the heat flux; $q0$ is the heat flux density in the area of meniscus; k is the coefficient describing the increase of total thermal resistance of the shell, the contact between the ingot and the mould surface in time.

However, in the equation (1), only v, z and L parameters are known for certain casting conditions. The rest parameters of the equation (1), $q0$, q_{mean} and k are unknown. The unknown variables can be determined with the aid of available industrial data. They include averaged values of temperature according to data from thermocouples in mould walls and the mean heat flux q_{mean} depending on casting speed.

The data from casting data sheets of the BOF production at PAO Severstal were collected and processed in order to determine the heat flux mean density value. 134 heats were processed in total. The value q_{mean} was calculated according to the formula (2):

$$q_{mean} = \rho_w g_w C_w \Delta T_w, \tag{2}$$

where ρ_w is water density, kg/m^3; g_w is water flow rate for mould cooling, m^3/s; C_w is water heat capacity, J/(kg·K); ΔT_w is mould input and output water temperature difference.

The parameters $q0$ and k of the equation cannot be determined directly through measurement, and the solution of the three-dimensional inverse problem of steady thermal conductivity in the Ansys 17.1 calculation complex with the optimization approach is used for their calculation.

The inverse problem is based on the following principle. The heat flux with distribution along the casting direction according to the equation (1) is applied to the three-dimensional model of the mould wall from the inside. From the other side – in areas of cooling channels – the boundary condition in the form of the contact heat exchange is applied. The coefficients in the equation (1) are based on the solution of the inverse problem so that the design temperatures in the areas of thermocouples comply with the experimental values.

The functions to be minimized when solving the inverse problem are expressed as follows:

$$F(q0,k)=\sqrt{\sum_{i=1}^{3}(T_i^{exp}-T_i^{calc}(q0,k))^2}\rightarrow \min\,, \tag{3}$$

where T_i^{exp}, T_i^{calc} are experimental temperatures in three locations (i = 1÷3) and their corresponding calculated temperatures received from solution of the three-dimensional direct problem of steady thermal conductivity.

The physical modelling was carried out for determination of the effective heat conductivity factor of the mould powder skull. A special unit, where the slag sample was located in the gap between the copper wall and the nichrome heater, was developed for these purposes. The skull effective heat conductivity factor was calculated on the basis of solution of the inverse problem of heat conductivity. The temperature values received when achieving stationary conditions, that is readings variability did not exceed ±2°C, were taken into consideration. The skull thickness was determined by a digital USB-microscope. The temperature values were recorded continuously during the experiment. The error of determination of temperature values by the MBA8 input analogue module is 0.5 %.

3. Study Results

The Figure 1 (a) presents linear dependencies of the mean density of the heat flux for mould powders Nos. 1 to 3 in relation to casting speed of peritectic steel grades. The highest density of the heat flux is observed for the mould powder No. 2. It is 200 and 400 kW/m^2 higher in average than for mould powders No. 1 and No. 3, respectively.

The Figure 1 (b) provides an example of temperature values according to the levels of thermocouples location in the mould. These dependencies were used for resolving the optimization problem.

The result of optimization is distribution of the heat flux density along the mould height. An example of heat flux density distribution for the peritectic range of products with use of the mould powder No. 3 is shown on Figure 2.

The values of heat flux density for different cross sections of the skull were calculated on the basis of received values of temperatures of the physical modelling of thermal and physical properties. Effective heat conductivity factors for a temperature range of 900 to 1,300°C depending on the slag skull thickness were calculated based on heat flux density values.

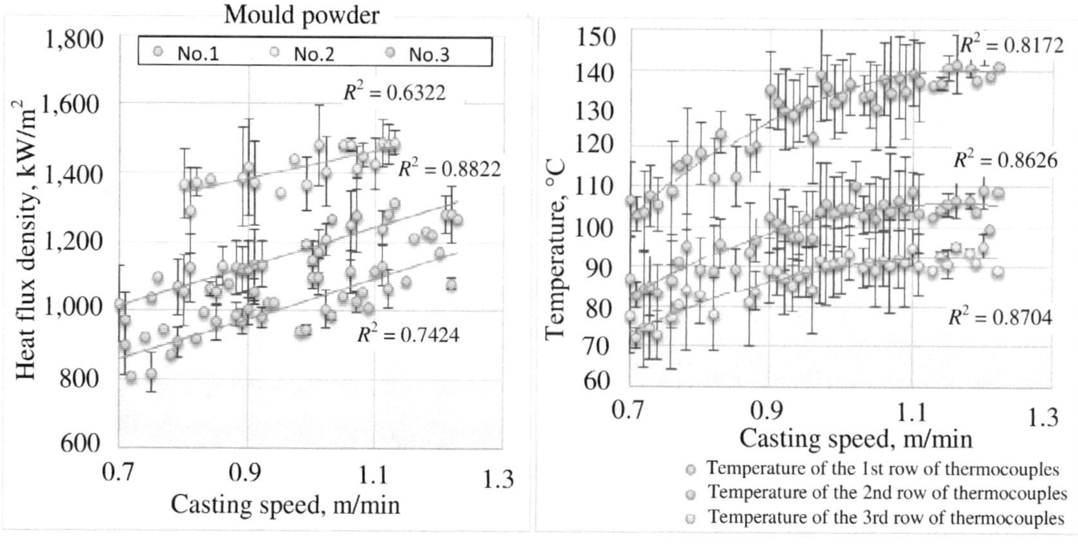

Figure 1. Measurement of heat flux mean density during casting of steel of peritectic grades (a) and readings of thermocouples (b) at casting of steel of peritectic grades with mould powder No. 3 depending on casting speed.

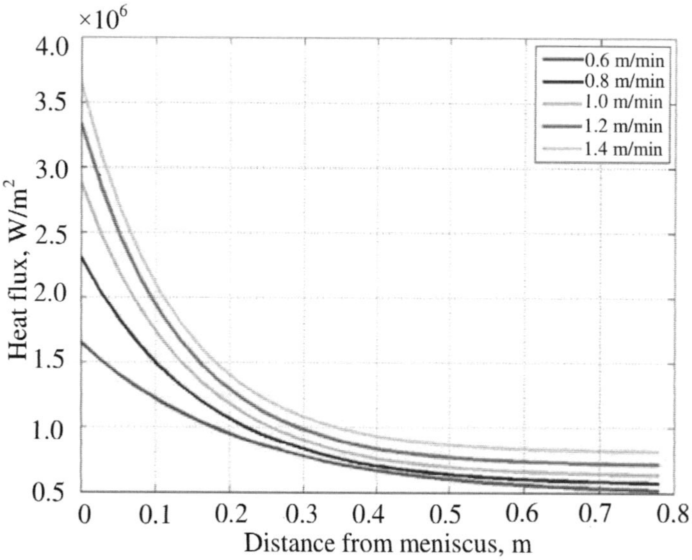

Figure 2. Distribution of heat flux density during casting of steel of peritectic grades with mould powder No. 3 for different speeds.

Then the mathematical processing of experimental data was conducted to obtain the dependence of the thermal resistance of the gas gap (resulting from molten slag shrinkage at the side of the mould) on the thickness of the mould powder skull. The following equation was considered for this purpose:

$$q = \frac{T_1 - T_2}{R_m} = \frac{T_2 - T_3}{R_c} = \frac{T_3 - T_{45}}{R_{int}}, \quad (4)$$

where T_2 is the temperature of mould powder crystallization; T_3 is the temperature at the border between the crystallized slag and the gas gap; R_m is the resistance of molten mould powder; R_c is the resistance of the crystallized mould powder; R_{int} is the resistance of the gas gap.

IOP Conf. Series: Materials Science and Engineering **287** (2017) 012001 doi:10.1088/1757-899X/287/1/012001

The equation (4) includes three independent equations, which clearly determine the resistance of the molten mould powder. The rest two equations contain four unknown variables: T_3, R_m, R_c, R_{int}, therefore, it becomes necessary to solve the optimization problem, which is formulated as follows:

$$f(R_c, R_{int}) = T_2 - qR_c - T_{45} - qR_{int} \rightarrow \min . \quad (5)$$

The solution of the optimization problem was obtained through standard numerical methods: the dependences R_{int} (d_{total}) and R_c (d_{total}) were determined. Figure 3 demonstrates dependencies of the gas gap resistance on total thickness of the slag skull according to the results of this study and the paper [7].

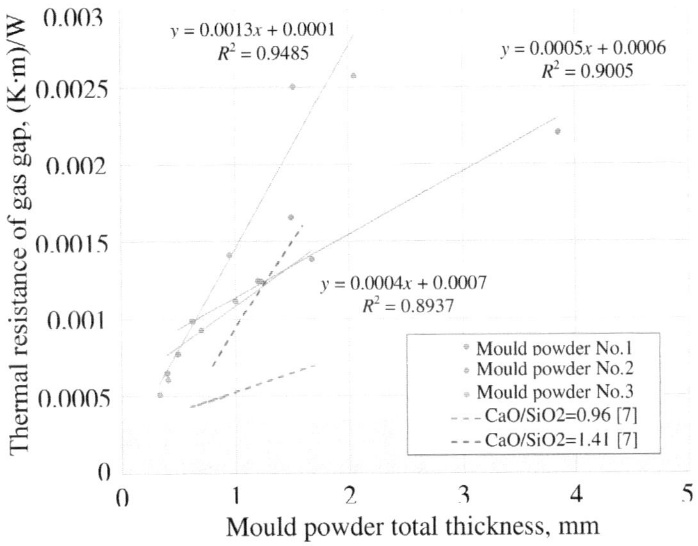

Figure 3. Thermal resistance of the gas gap on the surface of separation of the solid slag – copper wall depending on the mould powder thickness at 1,300°C.

Thus, when the heat flux distribution along the mould height and the value of thermal resistance of the gas gap between the mould wall and the solid slag layer are known, it is possible to calculate the mould powder skull behaviour in the gap using the mathematical modelling.

4. Mathematical Modelling

The results of the physical modelling were used as basic data for the mathematical model. The mathematical model determines distribution of liquid and solid slag, and the gas layer in the gap between the ingot and the mould, and it is based on the one-dimensional equation of steady thermal conductivity:

$$\frac{k_i}{d_i}(T_i - T_{i+1}) = q , \quad (6)$$

where k_i is the thermal conductivity factor of the ith layer; d_i is the thickness of the ith layer; T_i, and T_{i+1} are temperatures of layer boundaries; q is the heat flux, which is withdrawn by the mould from the continuous cast ingot.

When solving the problems of distribution of molten and crystallized slag, and the gas layer in the gap between the ingot and the mould, the equation (6) shall be written for three layers – for liquid and solid slag and the gas gap. In this case, the heat flux distribution and the ingot temperature along the casting direction are considered to be set.

Temperature distribution in the steel shell is based on the one-dimensional equation of the non-steady thermal conductivity:

$$p_{st}C_{st}\frac{dT}{dt} = k_{st}\frac{d^2T}{dx^2} , \quad (7)$$

where ρ_{st} is steel density; k_{st} is the steel thermal conductivity coefficient; T is steel shell temperature; t is time, sec; x are coordinates in the direction perpendicular to the casting direction, m.

The temperature distribution on the steel surface depending on the distance from the meniscus is determined out of the solution of the problem of temperature distribution in the steel shell (equation 7).

Solution of the three equation system consisting of three layers is achieved with introduction of some assumptions:

1) The gas gap is formed between crystallized slag and the mould wall, and its thickness is determined out of the physical modelling, the results of which are provided on Figure 3.

2) The three equation system shall be solved in sequence for z_i layer starting from the meniscus and ending with the lower part of the mould, where z_i is the distance of the ith layer from the meniscus.

The mathematical model basic data are presented below for determination of the slag layer thickness:

Thermal conductivity coefficient of mould powders Nos. 1-3 in the crystalline phase [8], W/(m·K)	0.5
Thermal conductivity coefficient of mould powders Nos. 1-3 in the liquid phase [8], W/(m·K)	1.7
Mould powder No.1 crystallization temperature, K	1,403
Mould powder No.2 crystallization temperature, K	1,298
Mould powder No.3 crystallization temperature, K	1,448
Steel thermal conductivity coefficient, W/(m·K)	33
Copper thermal conductivity coefficient, W/(m·K)	383
Gas gap thermal conductivity coefficient, W/(m·K)	0.023
Steel density, kg/m^3	7,800
Crystallization latent heat, J/kg	272,000
Steel solidus temperature, K	1,774
Steel liquidus temperature, K	1,779
Steel overheat temperature, K	1,788
Metal height in the mould, m	0.78
Distance from the mould wall to the thermocouple, m	0.02

Figure 4 presents an example of the gap profile between the ingot and the mould for the mould powder 2 depending on casting speed 0.6, 0.8, 1.0, 1.2 and 1.4 m/min, where 1 is the area of the ingot shell, 2 is the mould powder skull liquid layer, 3 is the intermediate layer, 4 is the solid layer, 5 is the gas gap between the mould wall and the skull solid layer, 6 is the mould wall.

Distribution of thicknesses of molten and crystallized mould powder was used for determination of dry and liquid friction.

According to the paper [15], dry friction $\tau_{s/s}$ (MPa) between the ingot and the skull solid layer can be determined by the equation:

$$\tau_{s/s} = \eta(\rho_{slag} g h_0 + \rho_{steel} g z), \quad (8)$$

where η is the friction factor between the ingot and the crystallized mould powder; ρ_{slag} is slag density; g is gravity acceleration; h_0 is height of slag liquid layer in the skull, and ρ_{steel} is steel density.

According to the same paper [15], it is possible to determine the liquid friction $\tau_{s/l}$ (MPa):

$$\tau_{s/l} = \eta_s \frac{(n+1)(v - V_m)}{d_l} - \frac{(\rho_{slag} - \rho_{steel}) g d_l}{n+1}, \quad (9)$$

where η_s is molten slag viscosity; n is a coefficient in the exponential correlation for viscosity; v is casting speed; V_m is mould oscillation speed, m/s; d_l is the thickness of molten mould powder.

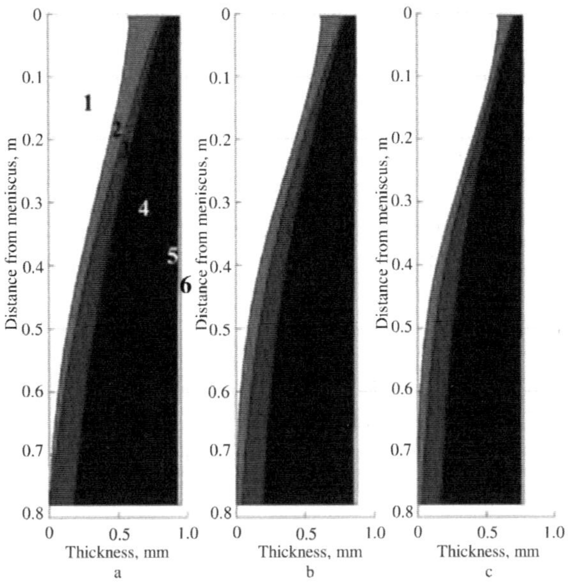

Figure 4. Profile of skull of mould powder No. 3 for casting speed (m/min):
(a) – 0.6; (b) – 0.8; (c) – 1.0.

Height (h_0, m) and thickness (d_l, m) of slag liquid layer in the skull was determined on the basis of results of the profile mathematical modelling (Figure 4). The other values are given below:

Coefficient of friction between the ingot and the crystallized mould powder [12]	0.15
Slag density, kg/m^3	3,000
Viscosity of molten slag of mould powder No.1, Pa·s	0.41
Viscosity of molten slag of mould powder No.2, Pa·s	0.18
Viscosity of molten slag of mould powder No.3, Pa·s	0.35
Coefficient in exponential correlation for viscosity	1.24
Casting speed, m/min	0.6 - 1.4
Maximal mould oscillation speed, m/s	-3.5

Friction stresses $\tau_{s/s}$ and $\tau_{s/l}$ were recalculated into the friction force or the ingot drawing force using the equilibrium integral equation. The liquid friction was calculated for the worst scenario, when the speed difference value in the equation (9) is maximal. Figure 5 shows calculated values of the friction force (ingot drawing force) between the ingot and the mould wall based on the equation (5) and represented as a dependence of the friction force on the coordinate z (for peritectic steel grades).

According to the analysis of Figure 5, the smallest value of the friction force appears with application of the mould powder No. 2. The rejection rate because of edge and transverse cracks is 0.36 and 0.89% relat. with application of the mould powder No. 2, 3.38 and 2.17 relat. with application of the mould powder No. 1, and 1.59 and 2.14% relat. with application of the mould powder No. 3, respectively. The largest friction forces appear with application of the mould powder No.3. Thus, the modelling results conform well to the industrial data on defects of continuously cast slabs.

5. Conclusions

Laboratory studies of thermal and physical characteristics of the slag skull of different mould powders were carried out. Dependences of thermal resistance of the gas gap between the mould wall and the skull solid layer were received.

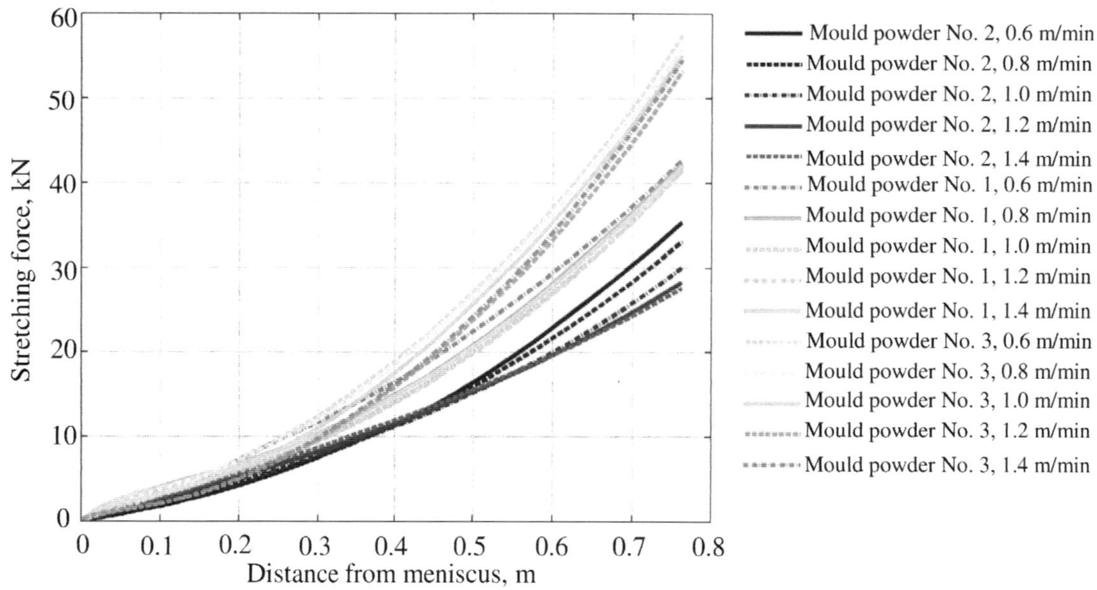

Figure 5. Calculated values of friction force (stretching force) for mould powders Nos. 1 to 3.

There was developed the mathematical model of heat transfer and slag solidification between the ingot and the mould, which allowed one to determine the thickness of the slag layer depending on the casting speed, steel grade, and mould powder type.

Friction forces and ingot drawing forces for different casting conditions were calculated based on the mathematical modelling of heat transfer in the steel shell and in the slag layer between the ingot and the mould. The received results conform to the industrial data on the slab defect rate.

References

[1] Anisimov K N, Kuklev A V, Lebedev I V and Toptygin A M 2014 *Prob. of Ferr. Metall. and Mat. Science* **2** 101–5
[2] Cho J W and Shibata H 2001 *Journal of Non-crystalline solids* **282** 110–17
[3] Zaytsev A I, Mogutnov B M and Shakhpazov E N 2008 *Physical chemistry of metallurgical slags* (Moscow: Intercontact science) p 352
[4] Anisimov K N, Longinov A M, Toptygin A M and Gusev M P 2016 *Steel* **7** 15–21
[5] Mahapatra R B, Sellers B T and Young I D 1988 *Steelmaking Conf. Proc.* 71 432
[6] Wang Q, Wang Y, Chi J, Xie B, He Y, Zhu B and Chen W 2014 *Material Science & Applied Chemistry*. **3** 65–8
[7] Cho J, Shibata H, Emi T and Suzuki M 1998 *ISIJ Int.* **38** 440–46
[8] Anisimov K N, Longinov A M, Gusev M P and Zarubin S V 2016 *Steel* **8** 32–7
[9] Lyakishev N P, Harutyunyan N A, Zaytsev A I, Zaitseva N E, Kalmykov K B, Leites A V and Shakhpazov E X 2005 *Metals* **3** pp 3–15
[10] Kwon O et al 1991 *Steelmaking Conf. Proc.* **74** pp 561–8
[11] Mills K 1991 *Steelmaking Conf. Proc.* **74** pp 121–29
[12] Meng Y, Thomas B G, Polycarpou A A, Prasad A and Henein H 2006 *Canadian Metallurgical Quarterly* **45** 79–94
[13] Makrushin A A, Zarubin S V, Aizin Y M 2006 *Steel* **5** 52–60
[14] Meng Y and Thomas B G 2003 *Metall. and Mater. Trans. B.* 34B pp 685–705.

[15] Wang X-D Kong L-W, Du F-M, Liu Y, Zang X-Y and Yao M 2014 *ISIJ International* **54** 2806–12.

IOP Conf. Series: Materials Science and Engineering **287** (2017) 012002 doi:10.1088/1757-899X/287/1/012002

Visualization of Process of Wheel Steel High Ingots Simulation

V I Bondarenko[1], V V Bodryaga[1], F V Nedopekin[1] and V V Belousov[1]

[1] Donetsk National University, 24 Universitetskaya Street, Donetsk, Ukraine

E-mail: vitbond@gmail.com

Abstract. The mathematical model for computation of formation of wheel steel high ingots has been formulated based on the generalized system of equations consisting of the Navier-Stokes equation, the turbulent heat and mass transfer equation and the continuity equation. It is suggested to use a pattern when designing software for simulation of hydrodynamic and thermophysical processes.
A software complex with friendly input and output data flows is provided for technologists of metallurgical production.

1. Introduction

Production of round ingots for making wheels or pipes is an important economic task, at the same time their continuous casting is impeded by high probability of crack formation. Therefore, the production is based on ingots casting into moulds.

Simulation of hydrodynamic and heat and mass transfer processes during casting and formation of metallurgical products is related to solution of complex non-linear multidimensional equations for computation of hydrodynamic and heat and mass transfer processes.

Nowadays, simulation systems are used in different industries, in particular in iron and steel industry, more and more often, which can be explained by apparent cost efficiency compared to pilot studies. The following classification features emphasize specific aspects of individual systems of computer-aided simulation [1] (Table 1).

In addition, there are also computational packages of overall profile, which, among other things, allow simulating transfer processes, such as ANSYS and COMSOL Multiphysics.

The expectations that modern integrated packages would easily implement complex models are not always met. Firstly, they are designed for calculation of streams of certain type, and any task different from a standard one requires significant preliminary actions: comparison of solutions received at different packages; evaluation of convergence and stability of the task, etc. It brings to nothing a key advantage of integrated packages – simplicity of mathematical models implementation. Secondly, if a package generates a finite-difference grid (finite elements) independently, then the major criteria is not the physics of streams, but the object geometrics; this could result in the generated grid failing to provide correct values in the area of large velocity gradients. Thirdly, these packages almost do not consider changes of metal state of matter.

IOP Conf. Series: Materials Science and Engineering **287** (2017) 012002 doi:10.1088/1757-899X/287/1/012002

Table 1. Specific Aspects of Individual Systems of Computer-Aided Simulation.

Name	Application		Functionality Level		Methods for Solving Mathematical Physics Equations			Orientation Towards Simulation	
	Special	General	Medium	High	Finite Differences (FD)	Finite Elements (FE)	Finite Volumes (FV)	Of Individual Physical Processes	Of Casting Technology Options in General
WinCast (FRG)	+		+			+		+	
MAGMASoft (FRG)	+			+	+				+
ProCast (USA, Switzerland)	+			+		+		+	
Poligon (Russia)	+		+			+		+	
LVMFlow (Russia)	+		+		+			+	
FLOW-3D (USA)		+		+			+	+	

Besides, a technologist, for whom the simulation results are eventually intended, shall be provided with a software product with friendly interface and input and output data flows convenient and clear to a production worker.

Therefore, development of application software packages (ASP) with comprehensive tools of computation visualization remains preferable at the moment for simulation of tasks of hydrodynamics and heat and mass transfer in metallurgical processes.

2. Mathematical Model

When designing the ASP, the authors suggest to use a pattern of Model-View-Controller (MVC), which is a scheme of partitioning of data of the application, the user interface and the control logic into three distinct components: a model, a view and a controller, in such a way that modification of each component can be done independently from others [2].

The model provides data and methods of working with them. In our case, the "Model" component in the basis includes libraries of classes and sub-programmes for computation of the given task of simulation of hydrodynamic and heat and mass transfer processes [3–13]. To describe processes of impulse, heat and mass transfer in molten metal, a mathematical model, which consists of equations of flow, continuity, heat and mass transfer, gaseous phase, turbulent kinetic energy (k), its dissipation rate (ε), and an equation for a solid phase part, is suggested ξ:

$$\frac{\partial \vec{V}}{\partial t} + (\vec{V}\nabla)\vec{V} = \nabla(\nu_{ef}\nabla\vec{V}) + \vec{g}\beta_T(T - T_0) + \vec{g}\beta_D(C - C_0) - \frac{1}{\rho}\nabla p; \quad (1)$$

$$\rho c\left[\frac{\partial T}{\partial t} + (\vec{V}\nabla)T\right] = \nabla(\lambda_{ef}\nabla T); \quad (2)$$

$$\frac{\partial C}{\partial t} + (\vec{V}\nabla)T = \nabla(D_{ef}\nabla C); \quad (3)$$

$$\nabla\vec{V} = 0; \quad (4)$$

$$\frac{\partial k}{\partial t} + (\vec{V}\nabla)k = \nabla(\nu_{ef}\nabla k) + G - \vec{g}\beta_T(T - T_0) - g\beta_T(C - C_0) - \varepsilon; \quad (5)$$

$$\frac{\partial \varepsilon}{\partial t} + (\vec{V}\nabla)\varepsilon = \nabla(\nu_{ef}\nabla\varepsilon) + C_1 \nu G - \frac{\varepsilon}{k}\frac{\nu_t}{\sigma_t}\vec{g}\beta_T(T - T_0) - \varepsilon^2 / k ; \qquad (6)$$

$$\xi(y) = 1 - \left(1 + \frac{c}{W}(T_l - T_s)\right)\left[1 - \left(\frac{T_l - T}{T_l - T_s}\right)^{2\beta}\right] + \frac{c}{W}(T_l - T_s), \qquad (7)$$

where $\vec{V}$ is melting speed, $\nu_{ef} = \nu_t + \nu$ is effective viscosity, ν is kinematic viscosity, $\nu_t = k^2/\varepsilon$ is turbulent viscosity; $\vec{g}$ is acceleration of gravity; β_T, β_D are hear and diffusion factors of volume expansion; T, T_0 is temperature current and initial; C is impurity concentration (carbon); C_0 is initial impurity concentration in the melt; ρ is melt density; p is pressure; c is coefficient of specific heat; $\lambda_{ef} = \lambda + \lambda_t$ is effective thermal conductivity, λ is molecular thermal conductivity and $\lambda_t = \nu_t / 0.9$ is turbulent thermal conductivity; $D_{ef} = (1-\xi)D + \xi D_s$ is effective diffusion coefficient, D is in liquid phases and D_s is in solid phases; G is a dissipative term; k is turbulent energy, ε is turbulent energy dissipation rate; C_1, σ_t is empirical turbulence coefficients [4]; W is latent heat of crystallization; T_l, T_s is temperature liquids and solidus.

The mathematical model is based on principles of macro-continual mechanics of the multiphase media and the theory of the mushy zone [4, 5]. The model addresses turbulence, gas capture by molten metal (during mould filling), and thermal and mixed convection. The system of equations (1) to (7) is supplemented by boundary conditions. It is important to note that the represented model of turbulence is not always justified [4]. It can be considerably simplified to an algebraic one for some types of stream.

3. Mathematical Model Implementation

The “Controller” component provides interaction between a user and the system, it controls and transmits data from the user to the system and vice versa, it uses the Model and the View for implementation of the required action [3].

The “View” component is a central theme of this paper. The View is responsible for receiving necessary data from the Model and their display to the user. The View can influence the condition of the Model, reporting it to the Model.

The scheme of the graphical user interface is shown on Figure 1.

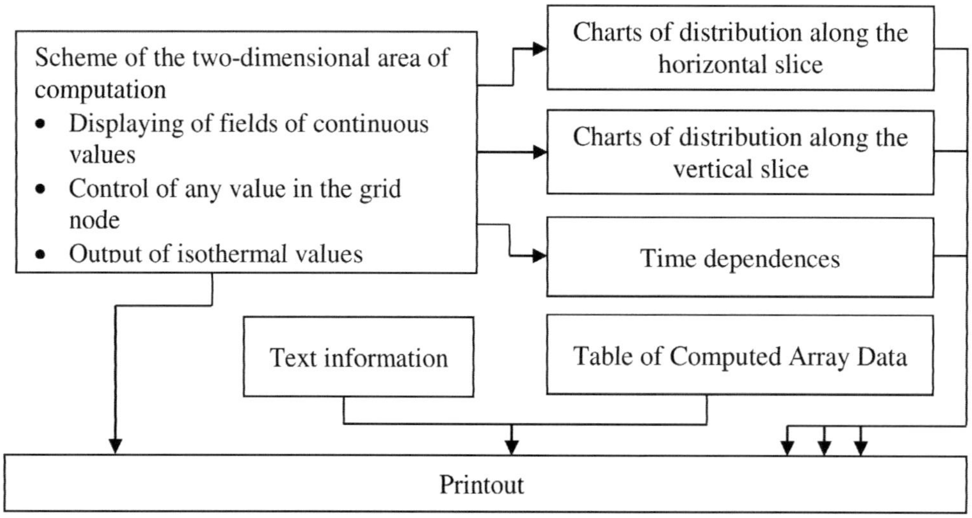

Figure 1. Graphical Interface.

The user can receive information in the following ways:

- By visualization of the area where computation is going on
- By curves of temperature, a share of the solid phase, a speed component, etc. at some horizontal or vertical slice
- By text view of computation arrays of temperature, a share of the solid phase, etc.
- Using values of selected data, such as shrinkage cavity.

Any window with information can be printed out. The time of computation, the per cent completion and prompt messages are displayed in the status line during operation.

The central section is a scheme of the two-dimensional area of computation. Generally, the task of simulation of metallurgical processes in moulds could be reduced to the two-dimensional axiosymmetrical model. The scheme of the computation area visualizes this model by colouring into different shades depending on the value of the selected field. Work with the fields of temperature, horizontal and vertical projections of the speed and a share of the solid phase is provided by default. But one can add any other fields representing two-dimensional arrays, for example, an array of local thermal stresses, etc. Interpretation of the computation area scheme is shown on Figure 2.

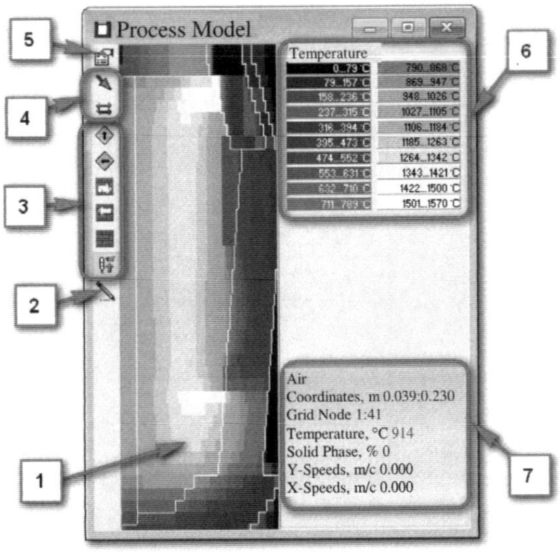

1 - coloured area, 2 - refresh button, 3 - field selection buttons,
4 - vector characteristics selection buttons, 5 - setting button,
6 - value range and colour shades setting panels,
7 - panel of information of computation grid current node.
Figure 2. Scheme of Computation Area.

The area 1 (Figure 2) represents the coloured scheme of some continuous value distribution, for example, temperature. A certain colour stands for each specific range of this value. Besides, different vector characteristics can be applied to this area, in particular, speed vectors. 18 options of value ranges and colours related to them are provided. A user can view and change these parameters through a set of panels 6 (Figure 2). Each program user can adjust the scheme view based on own preferences.

Switching of output fields is done by a set of selective buttons 3 (Figure 2).

If a mathematical model designer provides a special marking array, where the area geometry is described, contours of internal and external boundaries of a mould, a filler, etc. are laid on to the coloured area.

IOP Conf. Series: Materials Science and Engineering **287** (2017) 012002 doi:10.1088/1757-899X/287/1/012002

In addition, functions, which allow laying different isolines (functions of current, vortex structure, isotherms, etc.) on the area, and arrows indicating the speed vector direction are also developed. These functions are called by buttons 4 (Figure 2). The button 5 (Figure 2) calls the setting dialogue box.

Also the information is output from the current (under the mouse pointer) cell of the computation grid to the area 7 (Figure 7).

The graphic component for charting is a container of objects-graphs organized as a list element. One chart is a sample of Graphics class designed for description of the diagram. Features of this class are such characteristics as colour, line type, line thickness, etc. Data for a chart can be received as static points and as function-type variables. For the dynamic-type data, only the initial initialization is required. Later when a window is refreshed, the chart automatically gets updated, because it works with the function address and not with constant values passed to it. The other component characteristics are the coordinate grid (the amount of its cells, the line thickness, the colour, etc.), and the legend (the caption text and font settings). Besides, the evaluator of the mouse movement event is determined so that it could display the value of the coordinate grid under the cursor as an additional message, which is shown on the condition panel during performance of the program.

We discuss the use of visualization during simulation of high ingots (up to 3 m) of wheel steel. The Figure 3 demonstrates the screen of the running program. A metallurgical production technologist, who is a user of the software complex, can both evaluate at glance the processes occurring during simulation, and conduct a more detailed analysis.

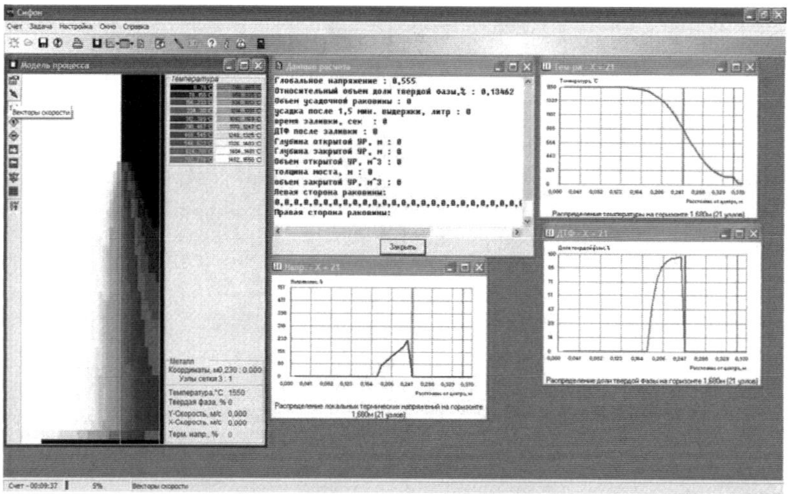

Figure 3. Screen of Running Application.

4. Mathematical Model Sufficiency

Experimental measurements of the liquid pool depth along the ingot axis during solidification (after completion of casting), the study of the shrinkage cavity depth in 8 ingots of KP-3 (Rus. "КП-2") steel, measurement of temperature of steel and iron moulds surfaces during casting and solidification of wheel steel ingots were conducted at OJSC "INTERPIPE NTRP" in order to evaluate the precision of the calculation results.

The shrinkage cavity depth at the head part of wheel steel ingots was determined on head parts randomly selected after cutting and breakage from 12 ingots of four heats (three from each heat).

The time of hot top filling of the wheel ingot according to the Technological Instruction NTZ-M-02-2005-1 (Rus. "НТЗ-М-02-2005-1") para. 7.4.7 was 2 to 3 min (the speed of 70 to 120 mm/min). The level of hot top filling with metal was lower than the upper edge of inserts approximately by the heat-insulation mixture layer thickness of 50 mm. The results demonstrate that the shrinkage cavity depth from the crown edge to the bottom of the open shrinkage cavity is 20 to 160 mm at the height of 130 to 190 mm from the crown edge to the ingot body.

In the numerical experiment, the shrinkage cavity depth was calculated under the condition of hot top filling with metal to the upper edge of the heat-insulation insert (mould butt end). In practice, it should be borne in mind that the metal pouring level is 40 to 50 mm lower than the upper edge of the slag mixture layer going to the hot top. During crystallization, metal level in the inserts goes down approximately by 50 mm, forming the level of the crown edge of the solid ingot top part [14].

The results of the numerical experiment of determination of open shrinkage cavity depth in ingots cast in iron and steel moulds during 10 to 18 minutes at the duration of filling of hot top insert 220 mm high of 120 to 160 seconds demonstrated that the shrinkage cavity depth in the entire study range is between 180 and 240 mm expectedly growing with the increase of the casting speed (Figure 4).

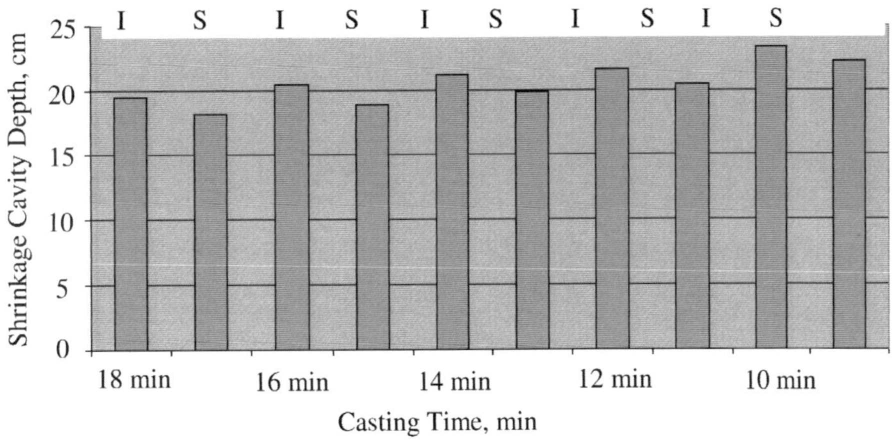

"I" and "S" are iron and steel moulds respectively

Figure 4. Calculation of Shrinkage Cavity Depth at Different Steel Casting Speed.

It should be noted that the shrinkage cavity depth received by numeric simulation is overstated, since the specific behaviour of the heat-insulation mixture, under which ingot casting is performed, when the metal is approaching the top part, is not taken into account.

Solidification of the ingot axial zone was evaluated by immersion of the metal rod into the ingot central part and determination of its immersion depth at different points in time.

The time of solidification of the ingot axial zone in the iron mould, according to simulation results, is about 90 minutes, and it coincides well with the pilot studies when extrapolating the curves.

Thus, the analysis of theoretical and pilot studies of wheel steel 8-billet ingot formation, conducted under conditions of the existing production of OJSC Interpipe NTRP, and the numerical experiment (Figure 5) demonstrated the appropriateness of the mathematical model, of the computational algorithm, and of the satisfactory correlation of the results obtained. The simulation uncertainty is 6%.

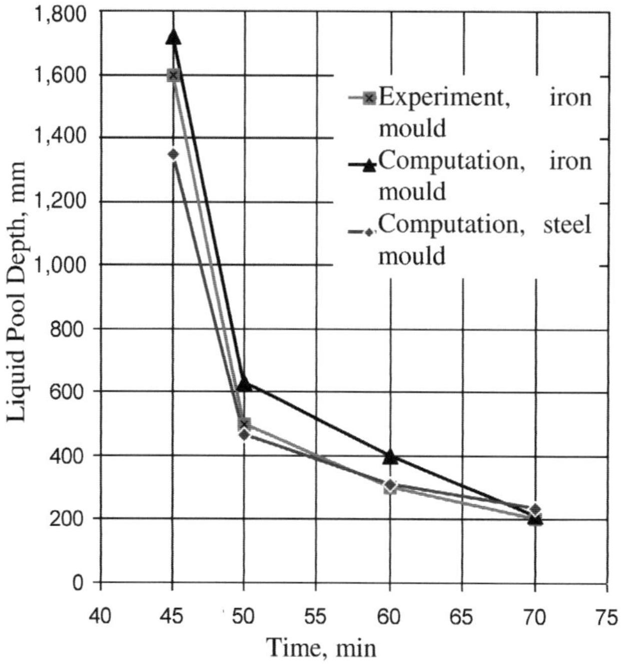

Figure 5. Comparison of Results of Industrial and Numeric Experiments of Determination of Liquid Pool Depth in Ingots Cast in Iron and Steel Moulds.

5. Conclusions

Visualization of results of the mathematical simulation of processes in wheel steel high ingots is demonstrated via examples of solution of the generalized system of equations of hydrodynamics and turbulent heat and mass transfer. The sufficiency of the mathematical model and the computational algorithm is proved. The Model-View-Controller pattern is used for designing the software for simulation of hydrodynamic and thermo-physical processes. A software complex with friendly interface and broad capabilities of simulation process visualization is provided for technologists of metallurgical production.

References

[1] Volnov I N 2007 The Systems of Computer-Aided Modeling – Status, Problems, and Prospects *Founder of Russia* **6** 14–17

[2] Fowler M, Rice D, Foemmel M, Hieatt E, Mee R and Stafford R 2002 *Patterns of Enterprise Application Architecture* (Addison-Wesley Professional) p 560

[3] Yu A Bazdyreva et al 2017 IOP Conf. Ser.: Mater. Sci. Eng. 192 012032

[4] Leibenzon V A, Kondratenko V M, Nedopekin F V, Belousov V V et al 2009 *Solidification of Metals and Metallic Compositions* (Kiev: Naukova Dumka) p. 446

[5] Anishchenko N F, Shablovsky V A, Belousov V V and Bondarenko V I 2006 *Metallurgical and Mining Industry* **3** 91–93

[6] Belousov V V, Babanin AY, Beskrovnaya M V et al 2013 *Metallurgist* **56** 933–37

[7] Shalapko Y I and Dobrzanski L A 2011 *Computer modeling of wheel steel ingots formation. Scientific basics of modern technologies: experience and prospects.* Monograph (Khmelnitsk: Khmelnitsky National University)

IOP Conf. Series: Materials Science and Engineering **287** (2017) 012002 doi:10.1088/1757-899X/287/1/012002

[8] Nedopekin F V, Belousov V V and Bondarenko V I 1997 *Artificial Intellect* **1–2** 86-90
[9] Nedopekin F V, Belousov V V, Kondratenko V M et al 2005 *Solidification of Metal Compositions: Production and Simulation* (Donetsk: South-East) p 231
[10] Bondarenko V I, Bilousov V V, Nedopekin F V and Shalapko J I 2015 *Archives of Foundry Engineering* **15** 13–6.
[11] Belousov V V, Bondarenko V I, Nedopekin F V, Bodryaga V V et al 2017 *Bulletin of Cherepovets State University* **1** 20–7
[12] Nedopekin F V, Belousov V V, Bodryaga V V, Bondarenko V I and Stepanov A T 2017 *Bulletin of Cherepovets State University* **1** 74-9
[13] Bodryaga V V, Nedopekin F V, Belousov V V and Okuneva T A 2017 *Bulletin of Cherepovets State University* **1** 28-32
[14] Bondarenko V I, Komarov V F, Belousov V V, Nedopekin F V et al 2013 *Mathematical support and Computer Technologies for Simulation of Hydrodynamic and Thermophysical Processes in Metallurgy*. Monograph (Donetsk: South-East) p 210

IOP Conf. Series: Materials Science and Engineering **287** (2017) 012003 doi:10.1088/1757-899X/287/1/012003

Ways to Improve the Quality of Die Steel 5KhNM

S V Efimov[1], O Yu Malykhina[1], A G Pavlova[1], V G Milyuts[2], V V Tsukanov[2] and V V Vikharev[2]

[1] OMZ-Special Steels LLC, St. Petersburg, RU-196650, Russia
[2] The Federal State Unitary EnterpriseCentral Research Institute of Structural Materials "Prometey" 49 Shpalernaya Street, St. Petersburg, RU-191015, Russia

E-mail: Semen.Efimov1@omzglobal.com

Abstract. There was performed an analysis of influence of the deoxidation technology, hydrogen content and high concentration of titanium in steel 5KhNM (Rus. "5XHM") on quality of die blanks, evaluated based on the results of the ultrasonic test.
The fractographic examinations of fractures and the X-ray microprobe analysis of chemical composition of non-metallic inclusions were conducted, the evaluation of macro- and micro-structure of a die blank with high titanium content was performed.
It is demonstrated that defects of dies from steel 5KhNM (Rus. "5XHM") are cracks from merged flakes and micro-flakes; in most cases large concentrations of sulphides appeared to be hydrogen collectors for formation of flakes and micro-flakes.

1. Introduction

Growth of unit capacity and technical characteristics of nuclear power plants, large power plants and chemical plants conditions the increase of weight and size of equipment parts. Die plates manufactured from large forgings of alloyed steels being in stable and steady demand on the domestic and international markets are used for manufacture of such products.

Manufacture of high-quality large forgings from forging ingots of large weight is by far a complex technical task, which is attributable to strong development of segregation processes, non-uniform distribution of non-metallic inclusions, flakes formation, physical heterogeneity inside an ingot and other deficiencies related to the process of solidification of large amount of metal. Moreover, increase of ingot weight results in enhanced development of defects.

One of common internal defects of forgings is flakes, which significantly deteriorate strength characteristics and plasticity of steel and structural strength of products made of such steels. According to classification of internal defects, suggested in paper [1], flakes are described as a defect of macrostructure in the form of thin curly cracks with a length of up to several dozens of millimetres located arbitrarily in different directions throughout the metal section except for the edge area. Flakes belong to the group of unacceptable defects.

Nowadays, it is considered proven [2, 3] that flakes are internal micro-voids filled with hydrogen. Nevertheless, so far there is no consensus on the mechanism of flakes formation, hydrogen "critical content" for their formation, and the role of nature and composition of non-metallic inclusions.

A. N. Morozov believed that the hydrogen threshold level ensuring absence of flakes in steel is 2.0 to 2.5 ppm [4]. The authors [5] also support this model of flakes formation [5].

On the other hand, papers [6-8] suggest the opinion that reduction of hydrogen content even significantly below this concentration does not guarantee absence of flakes in steel with very low sulphur content (less than 0.005%).

Therefore, known beliefs about conditions, under which flakes are formed, require clarification due to technology development providing extra low content of hydrogen, additives and non-metallic inclusions.

The authors of papers [9–13] expressed the hypothesis that flake germs are manganese sulphides, and it is related to their certain peculiarities, that is, they precipitate last of all during steel crystallisation, they fall into the shrinkage area, and their thermal expansion coefficient is much larger than the iron's one. When the ingot is cooling, manganese sulphides are decreasing in size more than the surrounding metal, which results in formations of cracks between manganese sulphides and metal, and as a consequence, these cracks presumably are the centres of flakes origination. The available empirical evidence indicates that manganese sulphides being "hydrogen traps" influence flakes origination. The amount of manganese sulphide particles decreases at very low sulphur concentration in steel, flakes are formed extremely intensively and it leads to noticeable growth of rejects because of flakes. This is due to the fact that because of low amount of manganese sulphides, the hydrogen concentration increases being attributable to each inclusion. This raises hydrogen partial pressure and enhances the possibility of cracks generation and development [6, 14].

Because of existing cases of titanium content in die steel 5KhNM (Rus. "5XHM") in the amount of 0.01-0.03% and more, which is significantly higher than the typical level, it is reasonable to view its influence on steel quality, since it is known that titanium content of more than 0.03% can result in deterioration of the quality of steel for castings because of possible formation of large sulphides and nitrides along the boundaries of grains [15, 16].

The analysis of literature sources, including those devoted to examination of quality of large forgings from alloyed steels, indicates that they almost do not contain any information on technological special aspects of manufacturing of forged products from large weight ingots.

At the same time, there remains an urgent need to develop optimal technological manufacturing methods preventing defects formation in large forgings.

This paper aims to study the influence of process parameters of melting and chemical composition on quality of forgings from steel 5KhNM (Rus. "5XHM").

2. Materials and Research Method

Stamp steel of 5KhNM (Rus. "5XHM") grade was smelted in the electric arc furnace EAF-120, and the semi-finished product was treated in the ladle refining installation.

The steel was teemed into forging ingots by the siphon method with protection of the metal stream from secondary oxidation by argon.

Two schemes of metal deoxidation during refining were used at steel smelting:

Scheme 1. Without aluminium (8 heats);

Scheme 2. With addition of aluminium before vacuum degassing (9 heats).

For trial heats, aluminium was introduced as flux-cored wire or as the aluminium chop over the slag.

The analysis of results of forgings production was carried out based on heat card data, ultrasonic test (UST) cards, forging data and heat treatment data.

There was determined the influence of the deoxidation scheme and hydrogen content in metal after vacuum treatment on quality of forgings.

The results of production of forgings from three other heats with titanium content in steel from 0.03 to 0.055%, which is 10 to 15 times higher than the regular level of its concentration in steel 5KhNM (Rus. "5XHM"), were analysed.

In view of high level of rejection of forgings from these heats, a sample taken from one of the rejected forgings from the heat under the reference number 3 was analysed in order to determine the type of indications determined during the ultrasonic testing.

Two templates were cut in longitudinal and transverse directions relative to the billet longitudinal axis. Then macro- and microsections were made from these templates, as well as samples for three-point bending were made out of the template in the longitudinal direction.

Macrostructure was determined by the chemical etching method in 15% water solution of ammonium persulphate.

Metal impurity by non-metallic inclusions was evaluated by comparison with GOST 1778-70 scales increased hundred times.

The fractographic examinations of fractures and the X-ray microprobe analysis of chemical composition of non-metallic inclusions were conducted by a scanning election microscope and an analysing add-on device. Fractures were received by the three-point bending method by disclosure of defects revealed at macrosections. The X-ray microprobe analysis of non-metallic inclusions was performed at microsections and from fractures.

3. Study Results

3.1. The Influence of Smelting Technology and Steel Chemical Composition on Quality of Forgings

The results of the ultrasonic test of forgings made from metal smelted according to two deoxidation schemes being studied are provided in Table 1.

Table 1. Influence of 5 KhNM (Rus. "5XHM") steel deoxidation technology on quality of forgings.

Technology Scheme	Deoxidation		Number of Heats, pcs.	Content in Steel, %		Number of Ingots, pc/t	Forgings Rejection Rate (UST), %
	Al	Ca		S	Al		
1	-	-	8	0.0055	0.004	17/600.35	**6.8**
2	+	-	9	0.0055	0.004	22/771.95	**6.3**

Comparison of the results of die steel production with aluminium and without it did not reveal any noticeable advantage of one way of metal deoxidation over another one.

Taking into account quite similar values of rejection rate of blanks for dies manufactured according to both ways of deoxidation, further study was performed using data from all heats without separation into methods.

The amount of rejected forgings, according to the UST, depends to a large extent on hydrogen content in metal after vacuum treatment. When hydrogen concentration was 0.8 ppm and less, defects in forgings were not found, and when this parameter was 0.9 ppm and higher, the rejection rate for forgings was 16.9% (Table 2).

Data from Table 2 suggest that the critical hydrogen content, ensuring absence of flakes in 5KhNM (Rus. "5XHM") steel forgings is about 0.5 ppm.

This conclusion is supported by the results of the previously performed research on combined influence of hydrogen and sulphur content in metal on flakes formation in large forgings [6].

At the same time, the experiments demonstrated that the actually achievable level of hydrogen concentration in the entire scope of manufactured die steel is 0.8 ppm and less under conditions of certain production.

Table 2. Influence of hydrogen content in steel 5KhNM (Rus. "5XHM") on quality of forgings.

Hydrogen Content in Metal After Vacuum Degassing, ppm	Number of Heats, pcs.	Number of Ingots, pc/t	Forgings Rejection Rate (UST), %
0.5	10	22/798.3	0
0.6 - 0.8	1	2/47.9	0
0.9 and more	6	15/526.1	16.9

The analysis of the results of production of die steel 5KhNM (Rus. "5XHM") with the titanium content of 0.03 to 0.055% (regular level of titanium concentration is 10 to 15 times lower and equals to 0.003 to 0.004%) demonstrated deterioration of forgings quality in this case. 40% of forgings produced from metal from three heats with higher content of titanium suffered defects, determined by the UST (Table 3).

Table 3. Results of UST of forgings manufactured from steel 5KhNM (Rus. "5XHM") with titanium content of 0.03 to 0.055%.

Heat Reference Number	Deoxidation		Content in Steel, %			Number of Ingots, pc.	Forgings Rejection Rate (UST), pc./%
	Al	Ca	S, %	Ti, %	H2, ppm		
1	-	-	0.008	0.03	1.1	2	-
2	-	-	0.002	0.03	0.5	3	1/33.3
3	-	-	0.005	0.055	1.4	5	3/60
TOTAL	-	-	-	-	-	10	4/40

The possible reason of high titanium content, not typical for steel 5KhNM (Rus. "5XHM"), is the fact that steel from these heats was subject to ladle treatment right after production of titanium-stabilized chrome nickel steel in these ladles. Therefore, it is recommended not to treat steel in ladles after production of chrome nickel steel with titanium in them in order to improve the quality of forgings from die steel.

3.1. Examination of Forging Steel with Higher Content of Titanium and Defects Determined by the UST

3.1.1. Macrostructure Examination. Flake-type thin curly cracks with the length of 3 to 12 mm parallel to the longitudinal axis of the blank (Figure 1) were observed after etching when studying the macrosections.

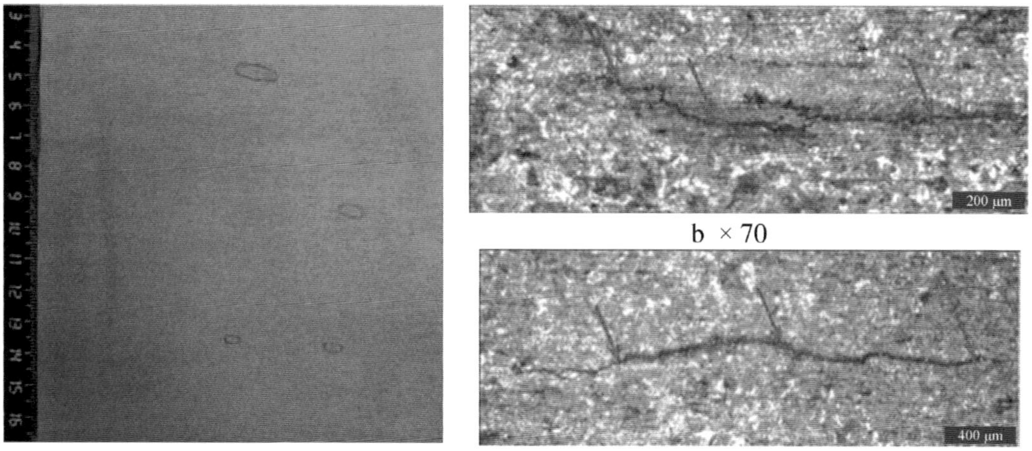

b × 70

a × 0.5 c × 30

Figure 1. Metal macrostructure (longitudinal direction)
(a) general view of macrosection; (b), (c) micro-cracks (pointed by the arrows).

3.1.1. Microstructure Examination and X-Ray Microprobe Analysis. Examination of longitudinal and transversal microsections without etching confirmed presence of flake-type thin curly cracks formed in the area of large concentrations of non-metallic inclusions of yellow-pink (predominantly) and grey colour (Figure 2). The X-ray microprobe analysis demonstrated that yellow-pink inclusions are titanium sulphides containing carbon, and grey inclusions are manganese sulphides.

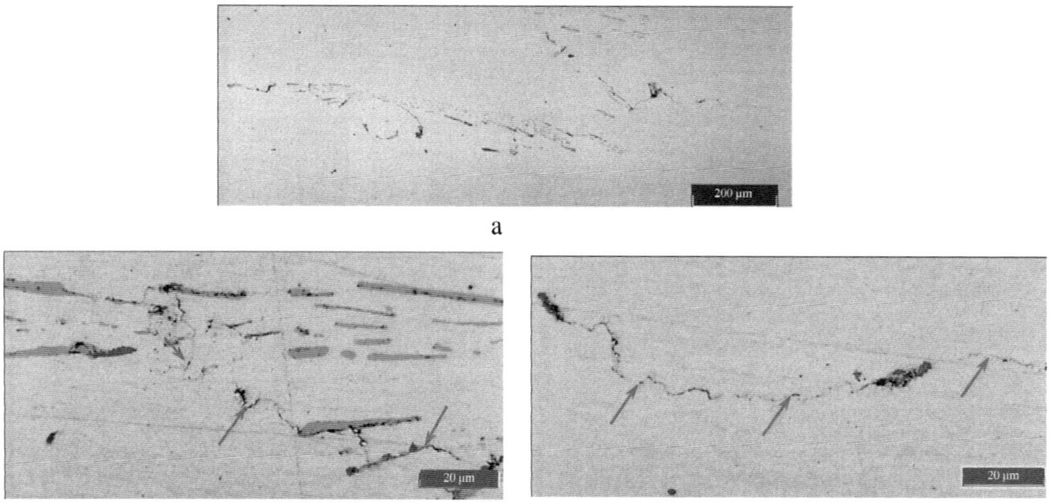

a

b c

Figure 2. Flake-type curly cracks formed in the area of large concentration of sulphides, the total length of the crack is ~1.5 mm, longitudinal direction: (a) general view, × 100; (b), (c) – fragments of the crack (pointed by arrows), high-power image, × 1,000.

Impurity of sample steel by sulphides corresponds to 4 points, by point-like nitrides to 4 points, and brittle silicates to 2 points of GOST 1778-70 scale (Figure 3). The nitrides are titanium nitrides/carbonitrides.

Steel impurity by titanium sulphides and titanium nitrides/carbonitrides, not typical for steel 5KhNM (Rus. "5XHM"), is significant.

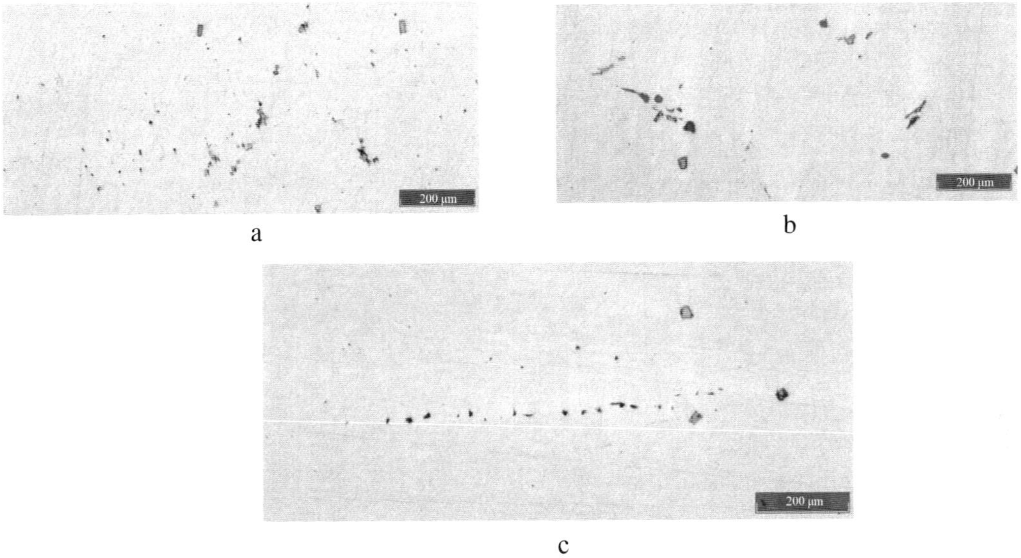

Figure 3. Non-metallic inclusions in sample steel, x100:
(a) point-type nitrides, transversal direction;
(b) sulphides and nitrides, transversal direction;
(c) brittle silicates and nitrides, longitudinal direction.

3.1.4. Fractographic Examination and X-Ray Microprobe Analysis The examination indicted that defects represent cracks formed from merged flakes/micro-flakes. Large concentrations of titanium sulphides also containing carbon (Figure 4) served as the hydrogen collector for them.

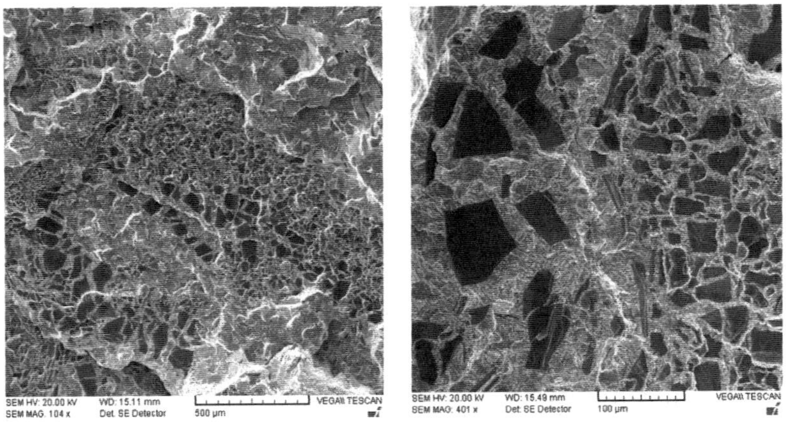

a × 50 b × 200

Figure 4. Photomicrographs of fracture surfaces.
(a), (b) – hydrogen collector – large concentrations of non-metallic inclusions.

4. Conclusion

The influence of the deoxidation technology, hydrogen content and higher concentration of titanium in steel on flakes formation in forgings from die steel 5KhNM (Rus. "5XHM") was studies.

It is demonstrated that critical hydrogen content in steel, which provides absence of flakes, is 0.5 ppm.

The risk of flakes formation significantly increases when titanium content in steel is 0.03% and more. In this case, large concentrations of titanium sulphides appear to be hydrogen collectors.

It is recommended not to treat die steel 5KhNM (Rus. "5XHM") in ladles right after manufacturing of chrome nickel steel stabilized by titanium in them, and vacuum degassing shall be performed providing hydrogen content in steel not greater than 0.8 ppm.

References

[1] Pravosudovich V V, Sokurenko V P, Danchenko V N et al 2006 *Defects of Steel Ingots and Rolled Products. Reference edition* (Moscow: Intermet Engineering) p 384
[2] Dubovo V Ya 1950 *Flakes in Steels* (Moscow: State Research Institute of Ferrous and Non-Ferrous Metallurgy) p 332
[3] Shtremel M A, Knyazev A A and Libenson A G 1982 *Kinetics of Flakes Growth. Physics of Metals and Metal Science* T 54 4 p 804
[4] Morozov A N 1968 *Hydrogen and Nitrogen in Steel* (Moscow: Metallurgy) p 281
[5] Yavoisky V I, Bliznyukova S A, Vishkarev A F et al 1979 *Inclusions and Gases in Steel* (Moscow: Metallurgy) p 271
[6] Vorobiev N I, Tokovoy O K and Mokrinsky A V 2003 *News of Higher Educational Institutions. Ferrous Metallurgy* 2 18–20
[7] Onishchenko A K 2007 *Die Forging* 1 8–11
[8] Fruehan R V 1997 *Proc. 13th Intemation Forgemasters Meeting (Korea)* vol 2 p 41–55
[9] Geld P V, Ryabov R A and Kodes E S 1979 *Hydrogen and Imperfections of Steel Structure* (Moscow: Metallurgy) p 219.
[10] Voronenko B I 1997 *Metal Science and Thermal Treatment of Metals* 11 12–18.
[11] Rosenfield A R, Shetty O K и Skidmore A J 1983 *Met Trans* 7-12 1934-37
[12] St. Pierre C R and Protasico S 1976 *Int Met Rev*. vol 21 269–79
[13] Ray G P, Jarman R A and Thomas J G N 1994 *J Mater Sci* vol 29 47–53
[14] Deryabin A A, Gorshenin I G, Matveev V V 2003 et al *Steel* 11 88–91
[15] Filippenkov A A 2002 *Steel Castings*: *Textbook*. (Yekaterinburg: Ural Department of Russian Academy of Sciences) p 276
[16] Korotich V I 2000 *Basics of Metallurgy: College Textbook* (Yekaterinburg: Ural State Technical University) p 392

IOP Conf. Series: Materials Science and Engineering **287** (2017) 012004 doi:10.1088/1757-899X/287/1/012004

New Engineering Solutions in Creation of Mini-BOF for Metallic Waste Recycling

S P Eronko[1], S M Gorbatyuk[2], E V Oshovskaya[1] and B I Starodubtsev[1]

[1] Donetsk National Technical University, 24 Universitetskaya Street, Donetsk, Ukraine
[2] National University of Science and Technology MISiS, 4 Leninsky prospekt, RU-119049, Moscow, Russia

E-mail: starbor45@gmai.com

Abstract. New engineering solutions used in design of the mini melting unit capable of recycling industrial and domestic metallic waste with high content of harmful impurities are provided. High efficiency of the process technology implemented with its use is achieved due to the possibility of the heat and mass transfer intensification in the molten metal bath, controlled charge into it of large amounts of reagents in lumps and in fines, and cut-off of remaining process slag during metal tapping into the teeming ladle.

The European Slag Association and the Institute of Processed Raw Materials Industrial Circulation were founded some time ago to solve the problems related to increase of the portion of the dormant scrap with high content of foreign matters and accumulation of environmentally harmful waste including ferrous and non-ferrous metals. According to these organizations, the total volume of slag utilization in developed countries have already exceeded 93%, and recycling of metal containing wastes allows the experts to produce annually useful goods worth more than $ 70 billion [1, 2].

Implementation of advanced technologies of industrial by-products utilization requires special melting facilities functioning within mini- and micro-plants, which best respond to market requirements of leading economies [3]. The positive experience of application of oxygen mini-cupolas and mini-blast furnaces for production of steel-making and commercial iron from coal and metallurgical sludge at such plants speaks for practicability of search for options to use small melting facilities capable to convert hot metal and metal-containing wastes with high content of harmful impurities into steel of acceptable quality without steel refiners in the processing chain.

As is well known, such units shall meet a set of requirements such as provision of thorough mixing of the molten bath, capability to control the feed of powdery reagents of varying composition into it, and prevention of releases of large quantities of the remaining slag into the receiving container during tapping [4].

According to the program of scientific and technological co-operation between the Mechanical Equipment of Iron and Steel Works department of the Donetsk National Technical University and the Process Equipment Engineering department of the National University of Science and Technology MISiS, the experts of these departments have been carrying out the work on creation of the basic oxygen mini-furnace complying with all requirements mentioned above over the past five years.

As a result of the performed comparative analysis of advantages and disadvantages of the known melting systems, a basic oxygen furnace with two rotation axes designed for the Kaldo process for producing steel from high-phosphorous iron at high consumption (40-50%) of metal scrap was used as a prototype when selecting the structural scheme of the unit being developed [5]. Due to necessity to eliminate several significant deficiencies identified in the BOF of this type, which have become the main reason for ceasing its industrial application in Europe, Japan, and the US, new technical solutions protected by Ukrainian and Russian patents were applied in design of the upgraded melting unit. The modified kinematic configuration of mechanisms of the converter body rotation about its longitudinal axis must first be noted. The upgraded mechanism (Figure 1) includes the bevel pinion shaft connected with the drive gearbox through the coupling and installed in bearing assemblies inside the cylindrical channel made in the trunnion body 2. The hollow trunnion is integral with the cover connected with the collar, where the shaft 1 with the bevel gear 4 set onto it and engaged with the gear shaft 3 is placed on frictionless bearings. The straight tooth gear mating with the toothed rim of the converter body is installed on the bottom end of the shaft [6].

Such design of the mechanism of the converter body rotation permitted allocation of its drive beside the corresponding trunnion of the mounting ring of the steel melting unit. It is much easier to mount, maintain and repair the rotating mechanism drive when it is permanently fixed on the work site [7].

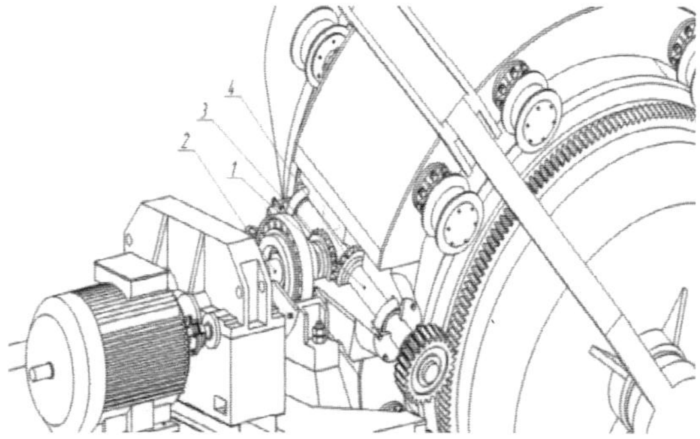

Figure 1. Design of the Converter Body Rotation Mechanism.

The converter was equipped with the system of molten metal stirring with an inert gas in order to improve the conditions of fast scrap melting, to increase the dephosphorization level along with reduction of burnt lime consumption, to reduce aluminium and ferromanganese consumption for killing steel, and to improve the refractory lining wear-resistance due to formation of less aggressive slag. This system allows, without using expensive automation tools, conducting the bottom stirring during unit body rotation in the pre-breakdown mode of the utility outflow considering the varying depth of lance submergence into the liquid bath. The offered stirring system (Figure 2, (a)) includes the main gas-feed pipe 1, the gas distribution unit 2 and several (3 – 4) multi-channel lances 3 mounted in the lining of the converter lower vessel cone 4. The gas-feed pipe, put through the channel in the driving trunnion of the converter, is laid up to the gas distribution unit 2 combined with the lower trunnion of the melting unit. The gas distribution unit (Figure 2, (b)) contains the hollow flanged trunnion 8 rigidly fixed by bolts 2 with nuts 3 to the bottom 1 of the converter body, and installed in roller bearings 7 and 10 located in the fixed support 11 connected by beams to the supporting trunnion ring of the melting unit. The shell 9 is mounted in the cavity of the trunnion 8 on slid fit. At some distance from its exposed face, the shell has a profiled slot 4 made at the lower semi-circle. The calculated width of the slot is maximal in the middle part, and it gradually narrows to the edges. The after end of the shell 9 is rigidly connected by the flange 12 to the body of the fixed support 11 and the cover 13 with the choke 14 connected with the pipe line 15. Through threaded holes are made at equal distance from each other in the body of the hollow trunnion. The longitudinal axes of the holes are in the trunnion's cross-section matching the axial semi-circle of the profiled slot 4 and the shell 9. The chokes 5 are screwed into the specified holes and connected with stirring lances by pipe lines 6.

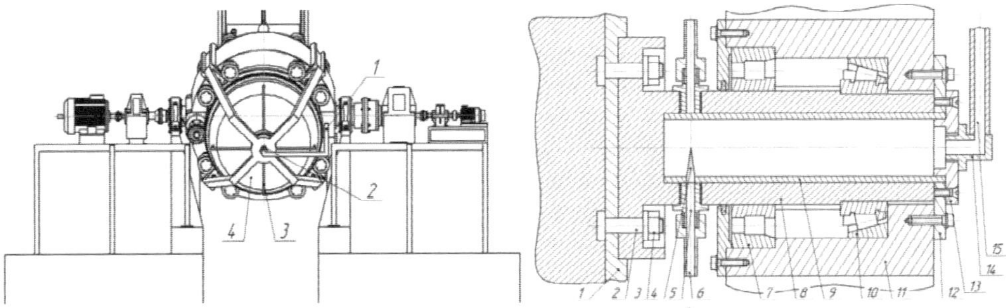

a b

Figure 2. Scheme of Allocation of Suggested Stirring System on Rotating Body Converter (a), Gas Distribution Unit Structural Scheme (b).

The principle of the gas distribution unit operation is the following. When the body 1 of the melting unit is rotating about its longitudinal axis, the trunnion 8 is turning in bearings 7 and 10 about the support 11 and the shell 9 rigidly connected to it. Gas delivered under pressure from the pipe line 5 through the choke 14 arrives to the cavity of the shell 9 and then through its profiled slot 4 to the channel of the pipe line, the choke of which currently coincides with it. Thus, gas is supplied by turns to each stirring lance in the automatic mode when they are below the molten metal level in the converter. Thanks to the changing slot widths, gas consumption is increased proportionally with the lance deepening into the molten metal achieving the maximum level in the lowest point of its immersion, and then it is gradually going down and its supply is interlocked when the lance is leaving the molten bath. Such operation of stirring lances provides efficient use of the inert gas and high efficiency of molten metal stirring thanks to elimination of so called breakdown mode of its outflow considering deepening of each lance [8].

In order to raise the efficiency of sulphur and phosphorus removal from the molten metal, the converter was equipped with the advanced system of top blowing by reagent powder in oxygen stream. The structural scheme of this system is provided on Figure 3. The system includes the mechanism of lance movement and the unit for metered feeding of the fine material. The mechanism of lance movement includes the trolley 5 equipped with two pairs of travelling rollers 4 installed on the inclined rails 3 (the angle of inclination is 30°). The lance trolley is moved by the screw drive mechanism consisting of the drive screw 18 rotated in the bearing assembly by the geared motor 17. The lead nut is located in a special housing attached to the bottom part of the trolley with the capability of self-alignment against the screw. The top part of the lance 1 is fixed on the supporting disc in the coupling 8, and together with the disk it can perform oscillating movement related to the trolley by means of the crank mechanism 6 within the elongated hole of the hood 2 rigidly attached to the trolley and provided with the socket for connection of the gas exhaust duct. In turn, the collar 8 together with the lance 1 held by it has the capability to turn about the supporting disc in vertical plane by a certain angle by means of the cam 7 installed on the output shaft of the geared motor 9 attached to the disc itself. Thus, thanks to the availability of the crank-lever mechanism and the cam mechanism, the lance provided with the nozzle in the lower part is enabled to perform oscillating movements relative to the trolley in two planes simultaneously when both mechanisms are on, or separately when the drive unit of one of them is off. Depending on the combination of amplitudes and oscillation frequency of the lance in two mutually perpendicular planes, the resulting trajectory of its nozzle movement is of different forms. During the steel-making process, it allows the operators to set such oscillation profile relative to the molten metal bath surface, at which the efficiency and factors of thermal performance of the process unit are improved the most. The lime powder metering device includes the sealed bin 13 fixed on four columns above the inclined rails and provided with a feeder located inside. This feeder consists of the geared motor 16 connected to the vertical hollow shaft 15 installed in the bearing assembly in the cavity of the protective pipe 14 and carrying the screw feed 12

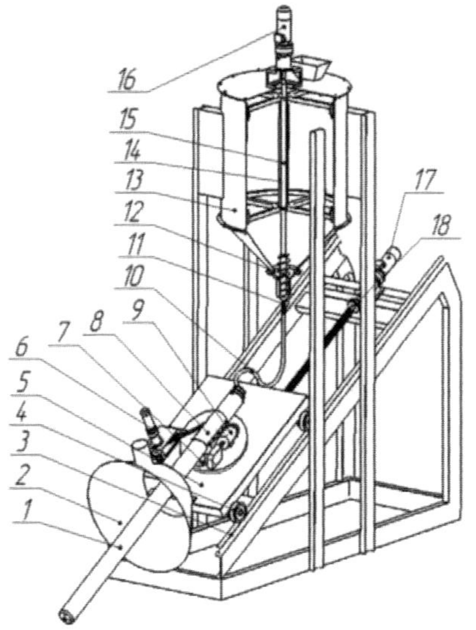

Figure 3. The Structure of the Suggested System of Top Stirring of the BOF Bath by Gas-Powder Mixtures.

on the shank. The lower part of the screw feed provided with the nozzle 11 is located in the mixing chamber connected with the choke of the oxygen lance 1 by a flexible pipe 10. When the gaseous oxygen is supplied to the cavity of the vertical shaft rotated by the geared motor at the permanent set angular velocity, the screw feed connected to it supplies the calculated amount of the lime powder into the mixing chamber, where its particles are captured by oxygen stream coming out of the nozzle and are taken to the lance duct by the pipe line mixing with the major amount of the oxidizing agent blown in to the converter bath [9].

In order to prevent the aggressive remaining slag from entering the teeming ladle, which might cause the excessive loss of deoxidizing agents and alloying additives, the upgraded mini-converter is equipped with the system of gas dynamic cut-off of the molten slag, schematically shown in Figure 4. It includes the hollow cylindrical rod 5, carrying on the front end the arc-shaped distribution box 8. In the wall of this distribution box, the slotted nozzles are present along the external perimeter. The rod is connected in a mobile manner with two pairs of rotary racks 3 and 7 pinned to supports 2 and 6 installed at different levels. A flexible gas-feed sleeve 4 is connected to the back of the rod, and a power pneumatic cylinder is used for rotation of the back pair of racks 3. The dimensions of racks 3 and 7 and the distance between their supports 2 and 6 are accepted based on the condition that they form the parallelogram mechanism together with the rod 5. Thanks to that, the rod is capable of parallel-plane movement in the vertical plane and it keeps the constant angle of inclination to the horizontal plane. Eventually, it provides the permanence of the set impinging angle of the flat gas jets, coming out of slotted nozzles of the distribution box 8, relative to the bath level of metal poured through the edge of the converter's nose into the ladle. The optimal impinging angle of gas jets provides effective holding of the molten slag in the converter due to pushing the slag aside from the converter nose under the influence of the dynamic head of gas flow [10].

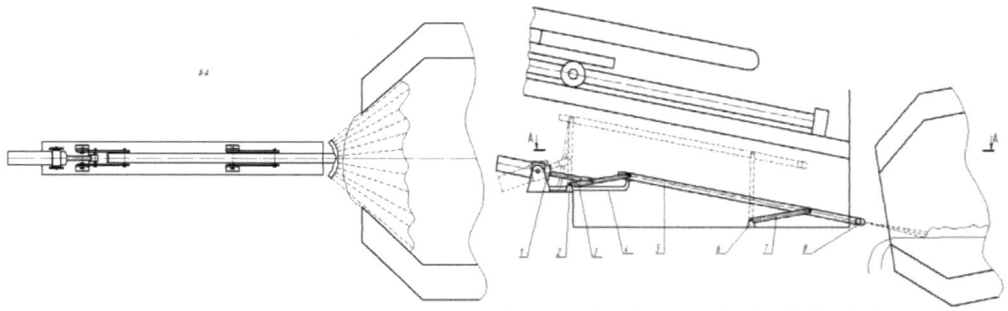

Figure 4. The Unit of Gas Dynamic Cut-Off of the Remaining Process Slag.

The thorough inspection of correctness of engineering solutions made during development of the upgraded design of the 10-tonnes basic oxygen furnace, performed on the current model of the unit, led to the conclusion that thanks to suggested innovations it has been possible to eliminate main deficiencies of the known equivalent [11]. Practical application of new developments will make possible the implementation of the efficient technology of steel production from low-quality charge materials without using additional facilities designed for refining of metal to be produced.

References

[1] Viner R 2011 *Metal Scrap* **2** p 9
[2] Ovchinnikov A M 2016 *Iron and Steel Industry: The Bulletin of the Institute "Chermetinformatsia"* **2** 96–100
[3] Smirnov A N, Safonov V M and Dorokhova L V 2005 *Metallurgical Mini-Plants* (Donetsk: OOO "Nord-Press") p 465
[4] Eronko S P and Klimovich N A 2016 *Iron and Steel Industry: The Bulletin of the Institute "Chermetinformatsia"* **5** 26–31
[5] Boichenko B M, Okhotsky V B and Kharlashin P S 2006 *The BOF Steel Production: Theory, Technology, Steel Quality, Unit Design, Materials Recycling and Environment* (Dnepropetrovsk: RVA "Dnepro-VAL") p 454
[6] Eronko S P, Fedyaev D I and Afaunova O V 2010 *Metallugical Processes and Equipment* **3** 40–46
[7] Sedush V Ya 2008 *Reliability, Repair, and Installation of Metallurgical Machines* (Donetsk: TOV "South-East, Ltd.") p 379
[8] Eronko S P, Oshovskaya E V and Starodubtsev B I 2017 *Iron and Steel Industry: The Bulletin of the Institute "Chermetinformatsia"* **4** 49–54
[9] Eronko S P, Oshovskaya E V and Starodubtsev B I 2017 *Iron and Steel Industry: The Bulletin of the Institute "Chermetinformatsia"* 7 75–81
[10] Eronko S P, Oshovskaya E V and Starodubtsev B I 2017 *Iron and Steel Industry: The Bulletin of the Institute "Chermetinformatsia"* **6** 48–54
[11] Artyukh V G 2015 *Basic Concepts of Metallurgical Machines Protection from Breakdown* (Mariupol: "University" Publishing Group) p 288

IOP Conf. Series: Materials Science and Engineering **287** (2017) 012005 doi:10.1088/1757-899X/287/1/012005

Analyzing Operating Behavior of Hot Mill Table Roll Drives using Statistical Methods for Current Values

P Z Kabakov[1], A V Kozhevnikov[1], I S Ilatovsky[2]

[1] Cherepovets State University, 5 Lunacharsky street, Cherepovets, RU-162600, Russia

[2] PAO Severstal, Mira Street 30, Cherepovets, RU-162608, Russia

E-mail: kabakovpz@mail.ru

Abstract. The article briefly describes failures of table rolls in Hot Rolling Mill 2000, PAO Severstal, and provides research results obtained by applying statistical methods. The statistical analysis showed a possibility of detecting deviations in roll behavior based on electric drive current changes, which cause failures. The statistical data analysis was employed to define roll critical behavior, which is to be checked using additional data. For this purpose, the research will be continued, roll operating behavior will be studied during the entire lifetime with due regard for deviations detected, comprehensive statistical analysis will be carried out, influence of rolling mill practices and roll failure predictability will be assessed.

1 Introduction

Maintaining process equipment in a good operating condition is among top priorities of any company. Scheduled maintenance and repair cannot always ensure trouble free operation. Steelmaking facilities (blast furnaces, BOFs, continuous casting machines, hot and cold mills etc.) are sophisticated engineering systems, therefore it is quite complicated to identify and consider all factors influencing failures. Different approaches are employed for failure cause analysis and maintaining fault free operation of equipment. As production process is rather complex, steelmaking facilities are highly automated. Automatic Process Control System (APCS) records data bulk, which can be used for research, search for dependencies, identification of factors causing failures. Therefore, statistical methods can be widely used for this purpose [1].

The article briefly describes failures of table rolls in Hot Rolling Mill 2000, PAO Severstal, and provides results of statistical investigation.

Most common roll failure is a seizure of mechanical components, which has a negative impact on neighboring rolls operation and increases probability of their failure. The increase in number of rolls to be repaired causes longer roll change time and lower mill productivity. Failures of run out table rolls can cause scratches and burrs on the strip.

As literature review shows, different approaches and methods [2–15] can be employed for diagnostics of electric drive mechanical components and failure cause analysis, including statistical analysis for changes in current and speed values. It should be noted that relatively few works [12–15] have been devoted to failure causes analysis of table rolls in hot strip mills. Besides, the works contain approaches and part of research results, and demonstrate how difficult and urgent it is to find solutions for roll failures.

2. Research and results

The research will consist of a few stages.

The first stage included data analysis for changed values of electric drive current and linear speed of rotating rolls and identification of deviations in rolls behavior related to the failure causes. The results of the first stage are presented in this article. It should be mentioned that a small number of rolls were used for analysis due to massive records per each roll, which is more than 4 million during 2-3 weeks of roll operation.

The second stage is to include collecting additional data regarding roll behavior deviations detected in the first stage during entire life and conducting analysis to prove their impact on failures.

The third stage is to include carrying out a statistical analysis of data bulk applying modern methods, assessing rolling practices impact on failures and developing a prediction model.

The fourth stage, which is the last one, is to include implementing software algorithms for table rolls diagnostics after maintenance and during operation.

The first stage included collecting data for statistical analysis regarding changed values of electric drive current and linear speed of rotating table rolls during relatively short periods of 3–4 week before failure. Rolls for analysis were selected by three with a faulty one in the middle. This selection pattern allows to compare operating conditions of the faulty roll with neighboring ones. Finally, 5 series, each including 3 rolls, were selected for analysis. For analysis, data was translated into more convenient form; statistical values were calculated: mean value and standard deviation, minimum and maximum, number of values exceeding 100 A and number of values of 200 A. Those values were calculated for one hour of roll operation. Data, recorded every 0.5 s, was reviewed for a detailed analysis of roll behavior. Comparative analysis of mean values and standard deviations was carried out for certain operating modes, when applicable.

The study included review of bar charts, correlation fields, time profiles of current and speed. Figures 1 and 2 show time profiles of current and linear speed. As Figure 1 shows, 400 operating hours of roll 22 saw numerous current peaks demonstrating load for electromechanical system. More detailed analysis of roll 22 operation revealed some patterns in current changes (Figure 2) and the fact that its operating behavior has lowest deviations against normal ones. Evaluation of roll 22 operating behavior confirmed that this roll could be used as a reference for comparison with operating behavior of other rolls and search for deviations.

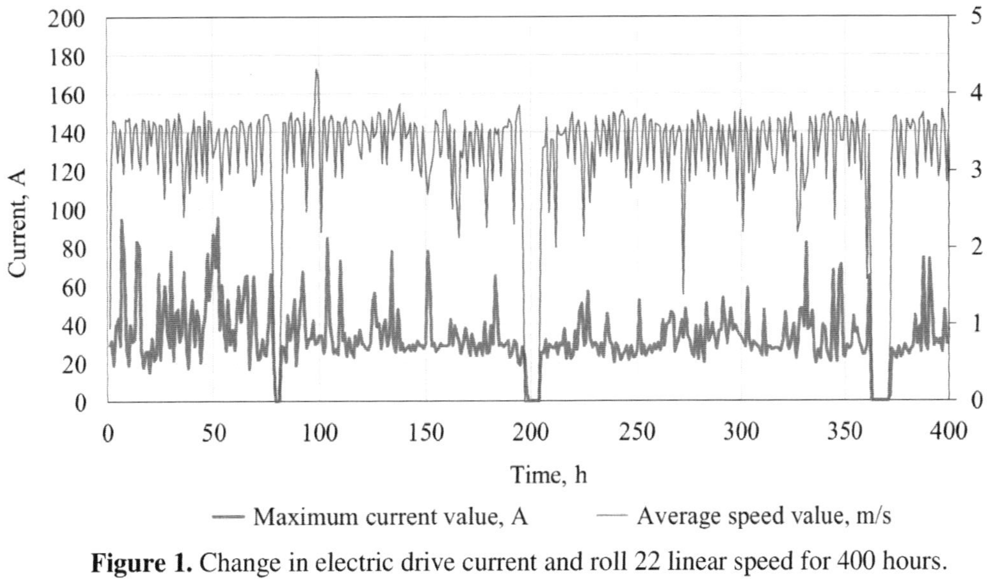

Figure 1. Change in electric drive current and roll 22 linear speed for 400 hours.

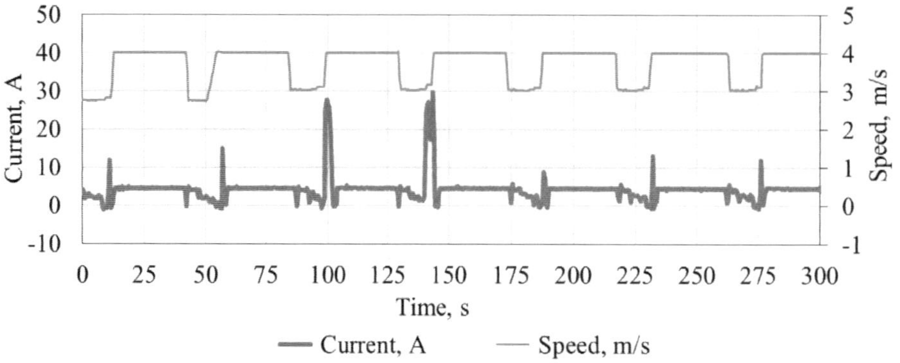

Figure 2. Change in roll 22 electric drive current for 300 s.

Analysis of other rolls operating behavior in terms of change in current showed different deviations for mean values, range, number of peak values exceeding 100 A and others. It should be noted, that among the most common deviations in roll behavior are peak values – current spikes from 100 to 200 A, which increase failure probability. Current spikes are seen in dynamic modes (slowdown) when load changes. Dynamic shocks in electromechanical system of the drive have a negative impact on its technical condition. Figure 3 shows roll 23 having numerous peak values. Roll 23 seized a few times within 400 hours of operation – during hour 175, 310 and 330.

Data was compiled in Table 1 for comparative analysis of roll operating behavior. As Table 1 shows all faulty rolls except roll 14 operated under conditions with peak values of 200 A.

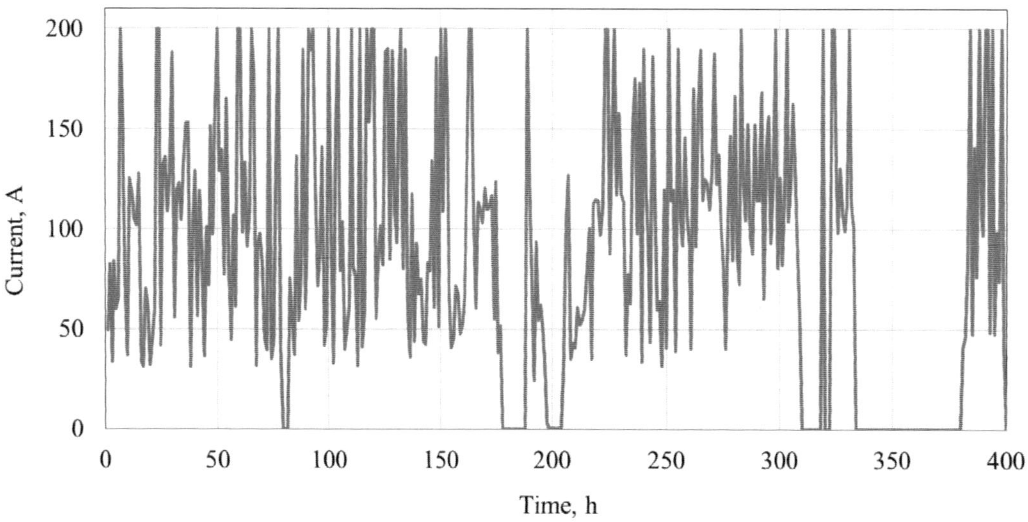

Figure 3. Change in current (highest values) of roll 23 electric drive for 400 hours.

Figure 4 shows time profile of current change for electric drive of roll 14, which failed due to the seizure of a bearing. As Figure 4 shows, average current values increased up to 25 A in second 506 (including increase in values range), then seizure occurred in second 1,108. At the same time, there were no obvious deviations in operating conditions of roll 14 before increase of the average value up to 25 A. Besides, there were no changes in operating conditions of neighboring rolls.

IOP Conf. Series: Materials Science and Engineering **287** (2017) 012005 doi:10.1088/1757-899X/287/1/012005

Table 1. Change in current of rolls' electric drive motors.

Series No.	Roll No.	Percent of rolls operating conditions, %					Note
		200 A	from 100 to 200 A	from 50 to 100 A	from 20 to 50 A	less than 20 A	
1	22	0.0000	0.0002	0.03	0.66	99.31	Operating
	23	0.0202	0.3874	0.64	8.49	90.47	Seized
	1	0.0000	0.0010	0.08	3.95	95.96	Operating
2	20	0.0006	0.0241	0.53	4.99	94.45	Operating
	21	0.0634	0.0450	0.57	16.90	82.41	Seized
	22	0.0005	0.0188	0.24	3.10	96.65	Operating
3	9	0.0031	0.0025	0.03	0.29	99.67	Operating
	10	0.0111	0.0107	0.43	35.95	63.60	Seized
	11	0.0049	0.0644	0.33	3.04	96.55	Operating
4	13	0.0000	0.0190	0.20	0.54	99.24	Operating
	14	0.0000	0.0048	0.21	3.59	96.19	Seized
	15	0.0000	0.0034	0.14	1.22	98.65	Operating
5	15	0.0000	0.0014	0.19	4.04	95.77	Operating
	16	0.0012	0.0045	0.38	4.15	95.46	Seized
	17	0.0000	0.0012	0.33	7.44	92.23	Operating

As correlation analysis showed, there was only a slight correlation in operating conditions of rolls neighboring the faulty one (the coefficient of determination is less than 0.2), that is a sudden change in electric drive current of one roll does not always cause current spikes of the neighboring roll. When no deviations occur, the coefficient of determination can reach 0.85. It should be mentioned, that these changes in roll (14) operating conditions indicate that not all failures can be detected beforehand. Table 1 shows that at the time of the study some of active rolls had current spikes up to 200 A, for instance, roll 11 (series 3). Therefore, data collection should be continued for checking an impact of detected deviations on failures. The second stage is to include collecting data regarding deviations of each roll during its entire life and performing analysis based on cumulative damage theory [11, 12].

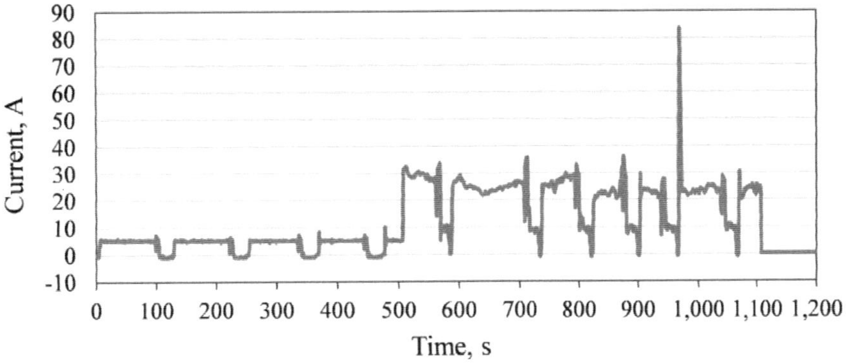

Figure 4. Change in current of roll 14 electric drive for 1,200 seconds before seizure.

IOP Conf. Series: Materials Science and Engineering **287** (2017) 012005 doi:10.1088/1757-899X/287/1/012005

3. Conclusion

In the first stage of the study statistical analysis was conducted for change in current of electric drives for intermediate table rolls in hot strip mill. It showed a detectability of deviations in roll operating conditions, which cause failures. Roll critical operating conditions were defined based on the statistical analysis results. In future collection of data on deviations detected will be continued and rolls operating conditions will be studied during their entire life.

References

[1] Kabakov P Z, Brechalov A S and Gainanov D N 2016 *Metallurgist* **8** 4–8
[2] Lukiyanov S I, Suspitsyn E S, Shvidchenko D V and others 2009 *News of Higher Educational Institutions* **1** 38–42
[3] Sidelnikov L G and Afanasyev D O 2013 *Bulletin of Perm State Research University. Geology. Oil, Gas and Mining* **7** 127–37
[4] Katsyn A N 2010 Casting and Metallurgy **1–2** (54) – (55) 250–54
[5] Verenev V V and Bolshakov V I 2006 *State of vibration diagnostics and monitoring of rolling mills equipment. Fundamental and Applied Problems of Ferrous Metallurgy: Collection of research papers* (Dnipropetrovsk.: Institute of Ferrous Metallurgy of Ukranian National Academy of Science) vol 12 p 267–281
[6] Krot P V 2011 *Mechanical drive issues: Collection of research papers* (Kharkov: National Technical University "Kharkov Polytechnical Institute") **29** p 68–81
[7] Shadsky G V, Yerzin O A and Rozhkov S V 2014 *Bulletin of Tula State University* **5** 144–50
[8] Kryukov O V 2013 *Automation and Remote Control* vol 74 **6** p. 1043–48
[9] Stepanov S Ye and Kryukov O V 2015 *Electrical Engineering: online academic journal* vol 2 **4** 53–61
[10] Yang Y and Bi Z 2014 *Proc. 2nd Int. Conf. on Advanced Cloud and Big Data (China)* p 213–19
[11] Zavoichinskaya E B and Kiiko I A 2004 *Introduction to Theory of Rigid Body Failure. Study guide* (Moscow: Moscow State University Publishing House) p 169
[12] Kozhevnikov A V, Ilatovsky I S and Solovyeva O I 2017 *Bulletin of Cherepovets State University* **1**(76) 33–39
[13] Shvidchenko N V and Lukiyanov S I 2010 *Bulletin of Tula State University* **3-3** 207-13
[14] Lukiyanov S I, Pishnograyev R S, Shvidchenko N V and others 2005 *Bulletin of G.I. Nosov Magnitogorsk State Technical University* 4 (12) 51–53
[15] Zhiltsov A P and Butyrsky D V 2016 *International Journal of Applied and Fundamental Research* **6-5** 827–29

IOP Conf. Series: Materials Science and Engineering **287** (2017) 012006 doi:10.1088/1757-899X/287/1/012006

Investigation of the Change in the Magnetic Properties of Die Steels in the Process of Manufacture and Operation of Die Tools

I I Karpukhin[1], A I Selishchev[2], A V Kornilova[1, 3], I M Idarmachev[4]

[1] ZAO Prochnost, 8A, Ryazansky prospect, Moscow, RU-109428, Russia
[2] AO KBAL im. L.N. Koshkina (Koshkin Engineering Center for Automated Lines), 21A, Prospect 50 let Oktyabrya, Klimovsk, RU-142184, Russia
[3] RUDN (People's Friendship University of Russia), 6, Miklukho-Maklaya str., Moscow, RU-117198, Russia
[4] FGOU FPO MGTU Stankin (Stankin Moscow State Technological University) 3A, Vadkovsky pereulok, Moscow, RU-127055, Russia

E-mail: karpukhin.iv@yandex.ru

Abstract. The paper presents the results of the investigation of the change in the coercive force in the materials of the die tool working parts during manufacture and operation of die tools; the paper offers possibilities of application of testing based on the coercive force.

The main problem stamping plants are facing is short lifetime of die tooling. Dies have complex shape and are subject to significant cyclic mechanical loads, and in case of hot stamping - also to temperature loads. Short service life of dies results in considerable production losses due to the equipment downtime caused by the change of stamps (or their working parts) and increased expenses for the equipment setup. Failures of die tools often cause emergencies on the shop floor and injuries to operating personnel. Die tooling lifetime is influenced by a large number of factors (equipment design and wear degree, process step load chart, design and running hours of a tool, etc.) and can vary significantly. A common technique to accurately determine a service life of a die tool has not been developed so far. The data given in the classical technical literature are empirical and approximate. There is no precise calculation method. The difficulty of estimating the tooling service life lies in the fact that the tool operation is almost always accompanied with a combined action of several destructive processes, which may both decelerate and significantly accelerate each other at different stages of the tool operation.

Therefore, it is required to identify a testable parameter to monitor the tool material condition in order to be able to take a tool out of service before it starts deteriorating. We propose to use the coercive force as such a parameter to monitor the die tooling condition [1]. The mechanism of change in the characteristics of the hysteresis magnetic loop under the influence of external or internal factors is based on the force and energy interactions of domain boundaries moving during remagnetization of a ferromagnetic with defects of a crystal lattice, with the contribution of electrons of incomplete *d*-shells of atoms to the generation of interatomic bond forces [2, 3], etc. Because of this, any loading of the interatomic bond forces caused by mechanical or thermal influences has an immediate impact on the magnetic properties of metal [4, 5]. This phenomenon has found use in non-destructive testing [6-8].

At present, diagnostics based on the use of the coercive force is exceptionally rare in die forging. This method is mainly used for testing hardness of the deforming tools, see papers [9, 10]. The purpose of this work is investigation of the change in the magnetic properties of die steels in the process of manufacture and operation of die tools and development of engineering foundations for diagnostic techniques based on the coercive force to assess the condition of cold sheet metal stamping dies and hot forging dies. Such techniques will make it possible: to determine the tool service life quickly and efficiently, to set the frequency of scheduled repairs, to mitigate economic risks for a company.

To check whether it is possible to monitor the condition of steels used for manufacture of cold stamping dies and hot forging dies with the use of magnetic methods we used the Schaeffler diagram [11], which represents the relation between the content of alloying elements in steel and its phase composition.

Alloying elements can be conditionally divided into two groups according to their influence on the phase composition, and, consequently, on the magnetic properties of steels. The first group includes elements which increase the magnetic permeability and reduce the coercive force. For this group, chromium equivalent m_{Cr} (%) is calculated using the formula [11]:

$$m_{Cr} = Cr + 1.5 \cdot Si + Mo + Nb \,, \tag{1}$$

where *Cr*, *Si*, *Mo* and *Nb* is percentage content of chromium, silicon, molybdenum, niobium, respectively. The second group includes elements which decrease the magnetic permeability and increase the coercive force. For this group, nickel equivalent m_{Ni} (%) is calculated using the formula [11]:

$$m_{Ni} = Ni + 0.5 \cdot Mn + 30 \cdot C \,, \tag{2}$$

where *Ni*, *Mn* and *C* is percentage content of nickel, manganese and carbon, respectively.
The calculation was performed for tool steels used for cold stamping (U10, U10A, U13, U13A, Kh12MF, 40Kh5MF, 27Kh2N2M1F, 8Kh4V3M3F2, 7KhG2VMF, KhVG, 9KhVG, 6KhS, 6KhVG, 4KhV2S) and hot forging (3Kh2V8F, 3Kh3M3F, 4Kh2V2MFS, 4Kh2V5MF, 4Kh2NMF, 4Kh3VMF, 4Kh4VMFS, 4Kh5V2FS, 4Kh5MF1S, 4Kh5MFS, 4KhMNFS, 4KhMFS, 5Kh2MNF, 5Kh3V3MFS, 5KhGM, 5KhNV, 5KhNVS, 5KhNM, 7Kh3, 8Kh3). The average values of chemical elements content specified in the corresponding regulatory documents were used for the calculation. Analysis of the results obtained with the use of (1) and (2) and plotted on the Schaeffler diagram showed that most tool steels have martensite-austenite structure, which means that magnetic methods can be used to monitor condition of such steels.

In order to check this conclusion we performed monitoring of 20 shearing dies for cold sheet metal stamping (to make a representative sample of the values received at the three plants where these dies are used in open crank presses). It should be noted that all the plants where we conducted experiments purchased tool steels from the same steel works (Zlatoustovsky Steel Works) and applied the same heat treatment process prescribed by the technical documentation. The coercive force H_c (A/m) was the main controlled parameter. The rated force of the presses varied from 125 to 500 kN; the tensile strength of the blank material varied from 450 to 930 MPa, the tensile strength of the female die material ranged from 1,235 to 2,550 MPa, the number of regrindings of the die being tested was from 0 to 2, the number of parts made with the die at the start of the testing varied from 0 (a new die) to 102,991. These factors were selected as the ones with most influence on the change in the coercive force in the die working parts and the ones that can be most reliably determined in the plant conditions. During the experiments we measured the coercive force in two mutually perpendicular directions at intervals equal to a certain number of loading cycles specified according to the capabilities of the real production facilities. Figure 1 shows one of the tested dies in the working area of the press. It was established that on repeated exposure to mechanical stress (without thermal cycling) the coercive force in tool steels increased. The coercive force gain was taken as a damage

IOP Conf. Series: Materials Science and Engineering **287** (2017) 012006 doi:10.1088/1757-899X/287/1/012006

indicator. The rate of its growth depends on the level of stress applied. The experiments show that an increase in stress by 13 MPa leads to an increase in the damage accumulation rate (by the growth of the coercive force) by a factor of 1.3.

Figure 1. Sequential cold sheet metal stamping die in the working area of the press.

The coercive force growth rate during cyclic loading in cold sheet metal stamping female dies (limiting the life of a die in most cases) is described by the following mathematical models:

$$\begin{aligned}\frac{dH_c}{dn} &= 0.85\cdot(7\cdot10^{-5}\cdot X_1^{\,2}-0.073\cdot X_1+19.105)\cdot(2\cdot10^{-6}\cdot X_2^{\,2}-0.008\cdot X_2+8.816)\times\\ &\times(0.520\cdot X_3^{\,2}-0.729\cdot X_3+0.651)\cdot(3\cdot10^{-10}\cdot X_4^2-3\cdot10^{-6}\cdot X_4+1.077)\times\\ &\times(-0.142\cdot X_5^{\,2}+0.652\cdot X_5+0{,}186)\cdot(7\cdot10^{-6}\cdot X_6^2-0.005\cdot X_6+2.321)\,;\end{aligned} \tag{3}$$

$$\begin{aligned}\frac{dH_c}{dn} &= 0.95\cdot(10^{-4}\cdot X_1^{\,2}-0.112\cdot X_1+29.160)\cdot(5\cdot10^{-6}\cdot X_2^{\,2}-0.017\cdot X_2+14.282)\times\\ &\times(-0.024\cdot X_3^{\,2}+0.384\cdot X_3+0.642)\cdot(2\cdot10^{-9}\cdot X_4^2-10^{-4}\cdot X_4+0.129)\times\\ &\times(-0.064\cdot X_5^{\,2}+0.379\cdot X_5+0.465)\cdot(5\cdot10^{-5}\cdot X_6^2-0.025\cdot X_6+4.648),\end{aligned} \tag{4}$$

where n is the number of the tool loading cycles; X_1 is the tensile strength of the blank material, MPa; X_2 is the tensile strength of the female die material, MPa; X_3 is the number of the tool regrindings; X_4 is the number of parts made with the die at the start of the testing; X_5 is the blank thickness, mm; X_6 is the rated force of the press, kN.

Equation (3) describes the change in the speed in the direction of measurement along the front edge of the press, equation (4) – perpendicular to the front edge of the press. Adequacy of models (3) and (4) to the experimental data is proved by the fact that the Theil coefficient is close to zero (0.065 for (3) and 0.124 for (4)). Based on the results of the research conducted, we developed a method and technique for determination of the life of tools operating under time-varying stresses (without contribution of thermal cycling towards the deterioration process) [12].

Until recently, the very possibility of using magnetic testing to estimate the service life of tools for hot forging has been questionable. At present, it is known that magnetic properties are sensitive to transformations that occur in metal during thermal cycling. This is confirmed by the results of our investigations of the magnetic properties of dies for hot forging and thixotropic stamping. We examined the coercive force of hot forging hammer dies of steel grades 5KhNV and 5KhNM for forgings of various configurations. Whenever possible, we took measurements at all stages of manufacture and operation of the die tools. In addition, we measured hardness in the same points with the use of a portative TEMP-4 hardness gauge. In parallel, the stamping process was calculated using the Deform-3D software according to the production process established for the tested dies and with

IOP Conf. Series: Materials Science and Engineering **287** (2017) 012006 doi:10.1088/1757-899X/287/1/012006

regard to the features of the applied equipment. The calculation gave us the stress and temperature distribution in the dies at the characteristic moments of stamping.

Next, we made samples of the measured values of the coercive force and calculated values of the maximum normal stresses and temperatures in the corresponding points of the die working cavity for each tested die condition. We used the stress and temperature values for the final step of calculation of the last stroke during stamping/sizing. For the die condition before stamping-sizing operations (at the stage of machining and heat treatment) we used the stress and temperature values typical for the same points of the die working cavity at the subsequent operation (stamping or sizing). Next, using Brandon's method [13] we approximated the coercive force values by the stress and temperature modules with the use of linear functions. This solution makes it possible to receive a clear picture of the coercive force distribution over the die surface and to monitor the coercive force changes depending on the loading level of the die working cavity areas during stamping operations.

Figure 2 (b) and (c) shows the obtained calculated dependences of the coercive force in the upper die for a "fork" forging. Measurements were taken for the die in a quenched condition before its operation and in a weakened condition after its operation and reworking. A 2-3-fold reduction in the coercive force between the measurements was recorded, with the coercive force being distributed more uniformly over all the points of the die working cavity after the processing. Figure 2 (d) shows a diagram for this die after quenching and stamping of 1,963 forgings. As can be seen from the diagram, there is a significant increase in the coercive force. In addition, we took measurements for the sizing

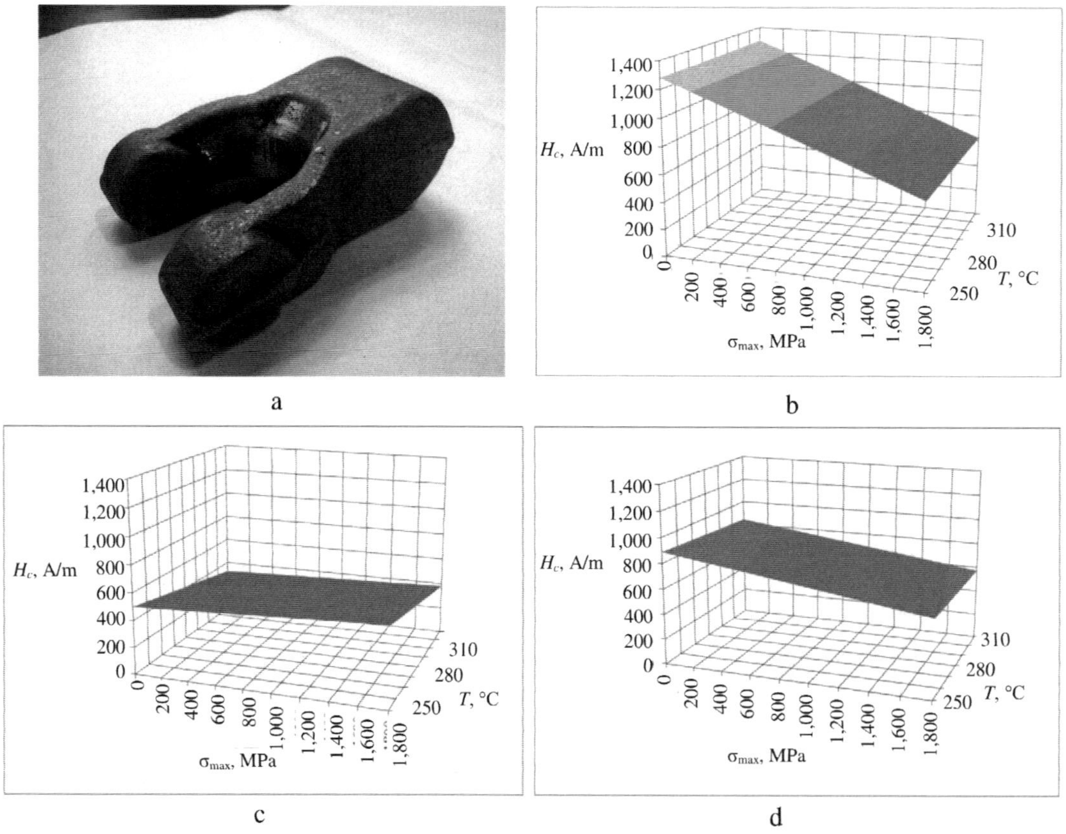

Figure 2. Dependences of the coercive force on temperature and stress for the upper die for a "fork" forging (a) in a condition: (b) before operation, (c) after stamping of 1,530 forgings and reworking, (d) after stamping of 1,963 forgings; die material – 5KhNV steel grade.

IOP Conf. Series: Materials Science and Engineering **287** (2017) 012006 doi:10.1088/1757-899X/287/1/012006

lower die for a "hoop" forging in a condition before quenching and in a condition after quenching and tempering with short time in operation. We recorded a 3-4-fold growth in the coercive force between the measurements. During annealing of steel a grain growth is observed, which is accompanied by a reduction in the total length of grain boundaries, which prevent displacement of domain boundaries, and, as a result, the coercive force decreases. Quenching, on the contrary, is accompanied by an increase in the length of grain boundaries due to formation of fine structure with precipitation of free carbon and carbides, which leads to the coercive force growth.

Figure 3 shows the surfaces of the sizing lower die for a "hoop" forging in a condition before quenching and in a condition after quenching and tempering with a small number of loading cycles. A 3-4-fold growth in the coercive force between the measurements was recorded. During annealing of steel a grain growth is observed, which is accompanied by a reduction in the total length of grain boundaries, which prevent displacement of domain boundaries, and, as a result, the coercive force decreases. Quenching, on the contrary, is accompanied by an increase in the length of grain boundaries due to formation of fine structure with precipitation of free carbon and carbides, which leads to the coercive force growth. The property of the coercive force to change during heat treatment is used in industry to control heat treatment quality and to determine steel hardness. In order to identify dependencies between the hardness and the coercive force of steel grades 5KhNV and 5KhNM, the approximate Brinell hardness *HB* for steel grade 5KhNV by the coercive force value H_c (A/m) may be determined using the formula:

$$HB = 0.0754 \cdot H_C + 259.3\,; \tag{4}$$

hardness *HB* for steel grade 5KhNM – using the formula:

$$HB = 0.095 \cdot H_C + 219.1\,. \tag{5}$$

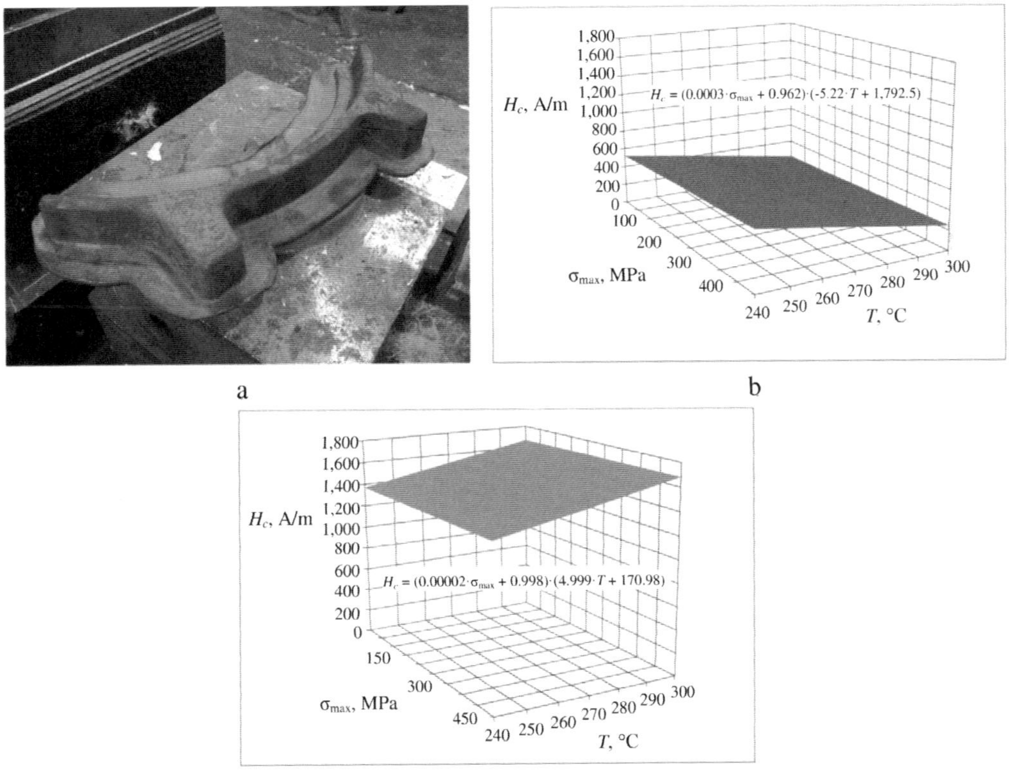

c

Figure 3. Dependences of the coercive force on temperature and stress for the sizing lower die for a "hoop" forging (a) in a condition: (b) before quenching, (c) after quenching and tempering and calibration of 90 forgings; die material – 5KhNM steel grade.

The obtained models are adequate to the experimental data by the Fisher's criterion with a significance level of 0.01. It is obvious from (4) and (5) that the hardness and the coercive force of steel grades 5KhNV and 5KhNM are almost in a direct relationship to each other.

Figure 4 shows a calculated diagram for the die for a "hoop" forging. This die cracked during heating. The low value of the coercive for steel grade 5KhNV suggests that the die has been in operation for a long time. The lowest values of the coercive force are recorded in the loaded areas.

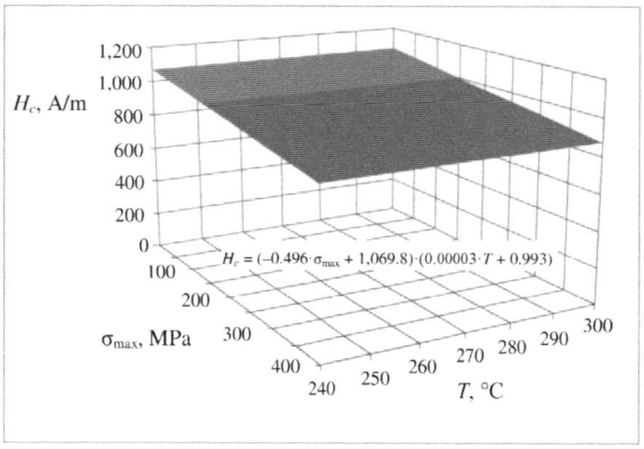

a

b

Figure 4. Dependence of the coercive force on temperature and stress (a) for the cracked lower die for a "hoop" forging (b); die material – 5KhNV steel grade.

Conclusions.

1. The experiments with cold sheet metal stamping dies have shown that the coercive force in all elements of dies in two mutually perpendicular directions of the coercive force measurement is unequal. The magnetic properties have shown that the tool damaging is different depending on the position of the die tool in the working area of the press. Therefore, there is a potential to increase the tool service life by means of optimization of the elastic line of a press table during process loading or through proper positioning of a die in the working area of an open press.
2. The conducted experiments and the calculations have proved that it is possible to develop an engineering technique for estimation of the life of cold sheet metal stamping tools.
3. The investigation of the magnetic properties of the hot forging dies material has shown the following: the coercive force increases in the die working cavity areas which are preheated to the recommended temperatures (200–300 °C), but which are not subject to significant mechanical and thermal loading during operation. The coercive force decreases in the die working cavity areas which are most loaded and heated. This decrease can be explained by the reduction in the length of grain boundaries due to the increase in grain size during heating, as well as by the increased efficiency of healing of the crystal lattice defects due to increased diffusion of atoms.
4. All the investigations conducted have shown that the coercive force increases during mechanical processing due to accumulation of the crystal lattice defects in the surface layer of the material.

References

[1] Kornilova A V, Idarmachev I M and Selishchev A I 2015 *Application of magnetic methods for testing of die tools* (Düsseldorf Germany: LAP LAMBERT Academic Publishing)
[2] Payne T O 1961 *Magnetic properties of small particles* (Moscow: Foreign Literature Publishing) p 198-225
[3] Levingston J D 1981 *J. Appl. Phys* **52** 2541-45
[4] Tikadzumi S 1987 *Physics of ferromagnetism. Magnetic characteristics and practical applications* (Moscow: Mir Publishing)

IOP Conf. Series: Materials Science and Engineering **287** (2017) 012006 doi:10.1088/1757-899X/287/1/012006

[5] Melguy M A 1984 *Magnetic testing of mechanical properties of steel* (Minsk: Nauka i tekhnika)
[6] Mikheev M N and Gorkunov E S 1993 *Magnetic methods of structural analysis and non-destructive testing* (Moscow: Nauka)
[7] Shcherbinin V E and Gorkunov E S 1996 *Magnetic quality control of materials* (Ekaterinburg: UO RAN, IFM)
[8] Bida G V 2006 *Magnetics properties of heat-strengthened steels and non-destructive examination of such steels* (Moscow: Marshrut)
[9] Vlasovets V M 2010 *Bulletin of Kharkov National Automobile and Highway University* **51** p 84
[10] Krutikova L A 2010 *Non-destructing testing and diagnostics* **3** p 21–27.
[11] Schaeffler A L 1949 *Metal Progress* vol 56 p 680
[12] Kuzminov A L and Karyshev A V 2011 *Health and safety* **5** p 35–38.
[13] Brandon DB 1959 *Developing mathematical models for computer control (ISA) Journal* **7** 70–73

IOP Conf. Series: Materials Science and Engineering **287** (2017) 012007 doi:10.1088/1757-899X/287/1/012007

The Study of Influence of Technology Factors and Equipment Design Parameters on Energy Consumption taking into Account Stochasticity of the Cold Rolling Process

A V Kozhevnikov[1], **N L Bolobanova**[1] **and I A Kozhevnikova**[1]

[1] Cherepovets State University, 5 Lunacharsky street, Cherepovets, RU-162600, Russia

E-mail: avk7777@bk.ru, kojevnikovaia@chsu.ru, bolobanovanl@chsu.ru

Abstract. Capabilities of simulation of the cold rolling process taking into account unsteadiness and stochasticity of its parameters are demonstrated. The analysis of the results of simulation of rolling power parameters with the decrease of the work roll diameter is carried out. An example of search for the most power efficient technology using a dynamic model is provided.

1. Introduction

Implementation of energy efficient technologies of thin steel strips rolling is of high importance for the increase of competitive ability of existing production facilities, and for the reduction of resource intensity and energy intensity of steel production. A significant aspect is the accelerated performance and implementation of inventions along with high reliability of the results, which requires improvement of methods of the technology parameters mathematical simulation considering equipment design parameters.

A method based on a deterministic description of the cold rolling process without taking into account process unsteadiness and stochasticity is widely used in practice of flat rolled products manufacturing [1 – 6].

Further, the overall development of the sheet steel rolling technology together with the stability of the process technology requires consideration of correlation of dynamic processes taking place in the driving line of the rolling stand, the roll set, and the deformation zone.

2. Model description

As shown in the paper [7], the following elements of the continuous rolling mill system shall be considered: plastic-deformation zones, hydraulic screw downs (HSD), and electromechanical systems (EMS) of stands, including automated electrical drives, mechanical transmissions, and work stands themselves (Figure 1). The following designations are accepted on Figure 1: u_i is the voltage at the motor's armature; ω_{ri} is the angular velocity of rotation of the work roll; M_{ri} is the torque required for the rolling process; υ_{i-1}, υ_i are the strip speeds at the exit from the $(i\text{-}1)^{\text{th}}$ and the ith stands; h_{i-1}, h_i are the strip thicknesses at the exit from the $(i\text{-}1)^{\text{th}}$ and the ith stands; N_{i-1}, N_i are the strip tension forces; P_i is the rolling force in a stand; g_i is the gap between work rolls; σ_{di-1}, σ_{di} are the strip deformation resistances in the $(i\text{-}1)^{\text{th}}$ and the ith stands; S_i is the lead factor.

The input coordinates of the electromechanical system are the voltages applied to armatures of electric motors of work stands, and the torque required for the rolling process; the output coordinate is the angular velocity of rotation of work rolls. In order to take into account influence of transient

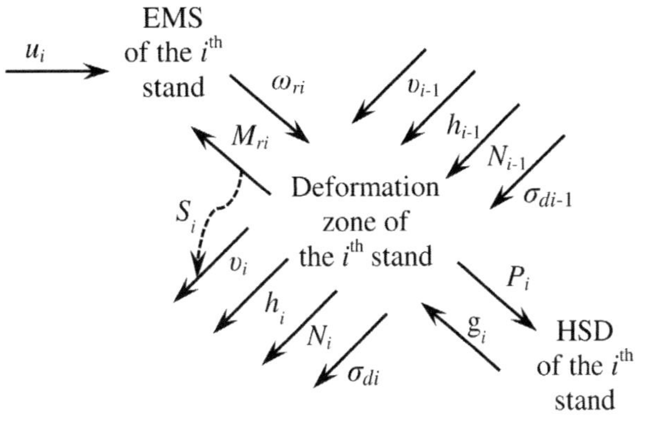

Figure 1. A Diagram of the Model of the Cold Rolling Process in the ith Stand of the Continuous Mill.

phenomena taking place in electric motors on the nature of dynamic loading in the electromechanical system of the stand, a description of a DC motor with independent excitation is introduced.

The deformation zone in the system of interaction of the strip with the work stand contains six input coordinates and six output coordinates. A mathematical description of the deformation zone of the ith work stand is based on the diagram where a strip is regarded as a thin elastoplastic body, and rolls are regarded as massive elastic bodies. Its detailed characteristic is provided in papers [8 – 10]. The impact upon the motor voltage is transferred to the entrance of the deformation zone through the speed of the rolls with the torque feedback.

All coordinates specified on Figure 1 are set considering the dynamics of the cold rolling process and stochasticity of its parameters. Parameters of the strip state like thickness, yield strength, factors of strengthening because of variations of heat chemical compositions are random values, which distribution is determined by the statistical simulation modelling.

When controlling strip thickness according to the Golovin-Sims method [11] the correlation between the deformation zone and the hydraulic screw down is performed through the rolling force and the roll gap. The gap is an input variable allowing to take into account external disturbances, e.g. the roll body whipping or controlling actions. A mathematical model of the hydraulic screw down coordinates control is described by the equations presented in the paper [7].

3. Results and analysis

The study of influence of technology parameters and design parameters of the revamped four-stand cold rolling mill of PAO Severstal on energy consumption was performed based on the developed dynamic model.

Design parameters of the roll set of the four-stand mill were changed by the renovation as follows: the diameter of work rolls was reduced by 20 mm, the diameter of back-up rolls was reduced by 50 mm, the length of the work roll body was increased by 420 mm, and the length of the back-up roll body was increased by 150 mm.

According to some studies [11 – 13], reduction of the work roll body diameter decreases capital expenditures and power consumption, however, reduction of the back-up roll diameter along with the increase of roll bodies has an adverse effect on rigidity of the roll system, and consequently, the accuracy of rolled strip with respect to off gauge and non-flatness is reduced. When studying design

IOP Conf. Series: Materials Science and Engineering **287** (2017) 012007 doi:10.1088/1757-899X/287/1/012007

parameters, a minimal diameter of work rolls of the renovated mill equal to 420 mm was taken as a basic case, and the diameter of work rolls equal to 350 mm with the possibility to increase the diameter of back-up rolls to save dimensions of the housing window was taken as an alternative case. This solution can provide the increase of rigidity of the roll set and the decrease of the vertical deflection of work rolls. A further decrease of the work roll diameter would require changes of the neck diameter of work rolls and back-up rolls that means that it would not allow using the existing bearings and chocks of the rolls, and would result in significant growth of the shear stress from transmitted torques.

The rolling mode of the 08ps ("08пс" in Russian) steel strip with the width of 1,850 mm, and the thickness of 0.9 mm from a work piece with the thickness of 3.2 mm, offered by a supplier of the renovated Mill 2100 equipment, which setting parameters are given in Table 1, was chosen for the study.

Table 1. Process Parameters of the Cold Rolling Mode for the Strip with the Thickness of 0.9 mm, and with the Width of 1,850 mm on the Four-Stand Mill 2100 of PAO Severstal.

Stand No.	υ_i (m/s)	h_{i-1} (mm)	h_i (mm)	ε_i (%)	N_{i-1} (kN)	N_i (kN)
1	8.27	3.2	1.995	37.66	20	75
2	12.2	1.995	1.35	32.33	75	95
3	17.6	1.35	0.937	30.59	95	100
4	18.3	0.937	0.9	3.95	100	35

ε_i is the individual relative strip reduction in the ith stand.

Calculated distributions of power parameters of the strip rolling process are provided in Table 2. According to the received data, the reduction of the work roll diameter results in the decrease of all power parameters and their variations. The reduction of the work roll diameter provides the decrease of the steel stress in the deformation zone because of the smaller length of the contact arc. When the work roll diameter is decreased from 420 mm to 350 mm, the rolling power is decreased by 10%. The comparison of distribution of the rolling process parameters suggests that the greater stability of power parameters of the rolling process is achieved with the decrease of the work roll diameter.

Table 2. Power Parameters of the Strip Rolling Process.
(Numerators are D_p = 420 mm, and Denominators are D_p = 350 mm).

Parameter	Stands			
	No. 1	No. 2	No. 3	No. 4
$\overline{P}$ (MN)	15.69 / 14.01	14.82 / 13.17	15.52 / 11.99	9.96 / 8.10
S_P (MN)	0.76 / 0.68	0.62 / 0.52	0.57 / 0.47	0.98 / 0.53
$\overline{N}$ (kW)	5,844 / 5,222	8,300 / 7,379	9,136 / 8,712	3,040 / 3,038
S_N (kW)	233 / 209	237 / 211	261 / 249	122 / 120

$\overline{P}$, $\overline{N}$ are average values of rolling forces and power; S_P, S_N are standard deviations of rolling forces and power.

The study of efficiency of strip rolling modes taking into account variations of the parameters starts with the procedure of simulation of yield strength and thickness values for each point of the work piece out of the user-set points. The simulation process is implemented using the procedure of generation of the random number according to the normal distribution law with the standard deviation of 10% for the yield strength, and 2% for the strip thickness. Reductions and interstand tensions range with account for restrictions on power and kinematic parameters; the rolling speed is recalculated on the basis of the law of volume constancy per second. After calculation according to the model, arrays of values characterizing the strip state and the rolling process parameters, which are estimated according to selected indices, are formed.

The values of reduction and tension for two process modes of rolling of the strip with the dimensions of 0.4×1,850 mm made of 08ps ("08пс" in Russian) steel are given in Tables 3 and 4 as an example of the methodology stated. In Case 1, reduction and tension are suggested by the equipment supplier, and in Case 2, the process parameters are determined by the simulation modelling of the rolling process. In both cases, the work piece thickness was 1.9 mm, and the original yield strength was 265 MPa, the rolling speed was 25 m/s.

Table 3. Reduction Modes.

Case No.	Individual Relative Reductions in Stands (%)			
	No. 1	No. 2	No. 3	No. 4
1	37	38	36	15
2	39	35	40	11

Table 4. Modes of Specific Tensions Broken down by Interstand Spaces.

Case No.	Specific Tension (MPa)				
	Before Stand 1	Space 1	Space 2	Space 3	After Stand 4
1	70	180	200	220	42
2	70	160	220	235	42

The tables suggest that in Case 2 the reduction is redistributed between stands — relative reductions in stands 2 and 4 were reduced, and in stands 1 and 3, on the contrary, increased, together with the growth of specific tensions in the 2nd and 3rd spaces and the decrease in the 1st space. The comparison of values of total power of the rolling mill motors (Figure 2) demonstrates that power reduction amounted to 1.17 MW during rolling according to the developed mode that means that expenses for the process technology are reduced by 5%.

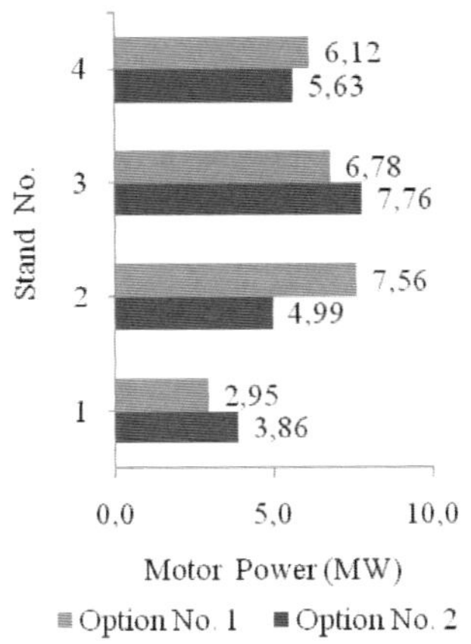

Figure 2. Comparison of Total Power of Rolling Mill Motors.

4. Conclusions

This paper presents rolling modes providing reduction of energy consumption due to correction of process parameters, and the results of the study of influence of the work roll diameter decrease on rolling force and power, received through the use of the dynamic model of the cold rolling process, which takes into account unsteadiness and stochasticity of its parameters.

The simulation results can be used for solving various tasks on optimization of design parameters of the rolling equipment and development of energy efficient production technologies of thin wide cold rolled strips.

References

[1] Garber E A and Kozhevnikova I A 2013 *Rolling theory* (Cherepovets: Cherepovets State University)
[2] Lee W H 2002 *J. of Mechanical Science and Technology* **10** 1296–02
[3] Roberts W L 1978 *Cold rolling of steel* (New York: Marcell Dekker Inc.)
[4] Vasilev Ya D 1995 *Engineering models and algorithms for calculation of cold rolling parameters* (Moscow: Metallurgiya Publ.)
[5] Khenzel A and Shpittel T 1982 *Calculation of energy and force parameters in metal-forming processes: Reference edition. Translated from German* (Moscow: Metallurgiya Publ.)
[6] Freshwater I J 1996 *Int. J. of Mechanical Sciences* **6** 633–48
[7] Kozhevnikov A V, Kozhevnikova I A and Bolobanova N L 2017 *Informatics and cybernetics* **2** 58–64
[8] Kozhevnikova I A 2010 *Development of the thin sheet rolling theory with a view to improve efficiency of wide-strip rolling mills* (Cherepovets: Cherepovets State University)
[9] Garber E A, Shadrunova I A, Traino A I and Yusupov V S 2002 *Russian Metallurgy* **4** 300–15.
[10] Garber E, Traino A, Kozhevnikova I 2009 *Flat-rolled steel processes: advanced technologies* ed V B Ginzburg (New York: CRC Press Taylor & Francis Group) pp 179–189.
[11] Tselikov A I, Polukhin P I, Grebennik V M et al 1988 *Machines and units of iron and steel plants* Volume 3 (Moscow: Metallurgiya Publ.)
[12] Garber E A 2007 *Manufacture of rolled products: Reference edition* Volume 1 (Moscow: Heat Engineer Publ.)
[13] Konovalov Yu V 2008 *Rolling Process Handbook.* Book 2 (Moscow: Heat Engineer Publ.)

Acknowledgements. This research was financially supported by the Ministry of Education and Science of Russian Federation within the framework of the project part of state task of Cherepovets State University (№ 11.3943.2017/ПЧ).

IOP Conf. Series: Materials Science and Engineering **287** (2017) 012008 doi:10.1088/1757-899X/287/1/012008

Competences of Engineers in the Iron and Steel Industry

A V Kozhevnikov[1]

[1] Cherepovets State University, 5 Lunacharsky street, Cherepovets, RU-162600, Russia

E-mail: avk7777@bk.ru

Abstract. The article presents the results of assessment of the professional performance of engineers working in the iron and steel industry. A competence-based profile of highly-qualified professionals has been built. The study of the competences of the iron and steel industry engineers has shown that their knowledge and skills may be mobilized to solve professional tasks.

1. Introduction

Along with investment opportunities, another key factor for successful upgrading and increasing competitiveness of the iron and steel industry products in the global market is the highly-qualified personnel capable of initiating and implementing innovations.

One of the internal factors of the Russian iron and steel industry that hinders its development is related to the emergence of the negative trend characterized by aggravating the problem of supply of highly-qualified personnel for iron and steel companies. Having in place a national competences system would make it possible to secure personnel selection and training, as well as timely identification of new skills requirements. In Russia, the national competences system is still under development.

Many authors both in Russia and abroad identify different types of competences, for example, papers [1 – 3] consider two groups of competences – *professional* competences and *social* competences, papers [4 – 8] – *special* competences and *general* competences or *hard* competences and *soft* competences. Most researches are aimed at the study of *social* (*general*, or *soft*) competences, with significantly fewer works focused on *professional* (*special*, or *hard*) competences. Various competence-based approaches to assessment of the high-technology companies personnel are being developed [6, 9, 10], levels of knowledge and skills required in the engineering labor market are being evaluated, benchmark studies of engineering competences demand and supply in Russia and the EU countries are underway [4]. The European Centre for the Development of Vocational Training (Cedefop) [11] conducts broad-scale researches related to competences assessment and forecasting personnel competences demand in Europe.

2. Methodology

For the purpose of efficient implementation of HR policy in the iron and steel industry, which technical-and-economic level determines the development strategy for the whole economy, of vital importance are the tasks of selecting and training highly-qualified personnel with regard to the demand for new skills for implementation of technological innovations.

In the absence of a system solution for assessment of professional performance of engineering personnel it is not possible to simply apply the existing competence-based approaches. Individual mechanisms which are already in use [12] – getting feedback from employers, identifying a mismatch between demand and supply in the labour market – may be combined to identify and assess professional competences. The use of these tools allows identifying a relation between the obtained qualification and the job profile, additional training needs of qualified specialists to succeed in the labor market, frequency of using various forms of continuing vocational training by graduates, existing demand for additional knowledge and readiness to get such knowledge. It is important to perform competences examination. This is done by conducting a questionnaire survey among groups of respondents with the use of the following three main approaches [13]: 1) analysis of the answers to indirect questions about the value of a university degree, professional experience, level of satisfaction by the received professional education; 2) competence measurement by descriptors describing the relevant routines (professional duties, main activities or specific activities at the work place); 3) direct assessment (self-assessment) of the competences a respondent possesses or of the required competences.

3. Results and analysis

The above methods and approaches were tried out for assessment of professional competences of the engineering personnel employed at Severstal Russian Steel division of Severstal. Competences assessment involved analysis of skill levels and personal properties of respondents. We used the tools that were applied in the European research “REFLEX” [14] and in paper [4], including self-assessment of competences by the respondents against the proposed list and on a common scale. The developed list of competences covered by the research is given in Figure 1.

In order to match the existing skills with the required skills the respondents were asked the following questions: “How would you characterize your level of the following knowledge and skills?”, “What level of the mentioned knowledge and skills does your job require?”. The self-assessment procedure covered 24 skills and was carried out based on the same 7-point scales. The following groups of competences were analyzed: a) instrumental competences including professional knowledge and communicative skills; b) interpersonal skills reflecting functional flexibility and ability to work in a team; c) system competences reflecting ability to mobilize the existing resources, ability to perceive the new, managerial skills. Focus was placed on the analysis of general (*soft*) competences reflecting the level of proficiency in the use of social, communicative and managerial tools and lifelong learning ability. It should be noted that *soft* skills are inseparable from *hard* skills, and are also *professional* skills, while not relating to one profession only.

The competence-based profile of the highly-qualified personnel plotted on the basis of the average score of the sample group is shown in Figure 2. The results of the survey show that the engineers give high scores to almost all of their knowledge and skills being evaluated: the scores given by the respondents were seldom lower than 5 on the 7-point scale. A similar trend is observed when it comes to the required competences. The required level of competences exceeds the existing one, it was only in a few cases that the respondents stated that their level of skill was higher than required. The biggest gaps between the existing and the required level of competence are observed for the ability to talk on professional topics in a foreign language and the ability to sell products or services.

The results of the study of competences of the highly-qualified personnel in the iron and steel industry show that the engineers are active in their professional lives, focused on their professional development and career advancement, and committed to work in the iron and steel industry.

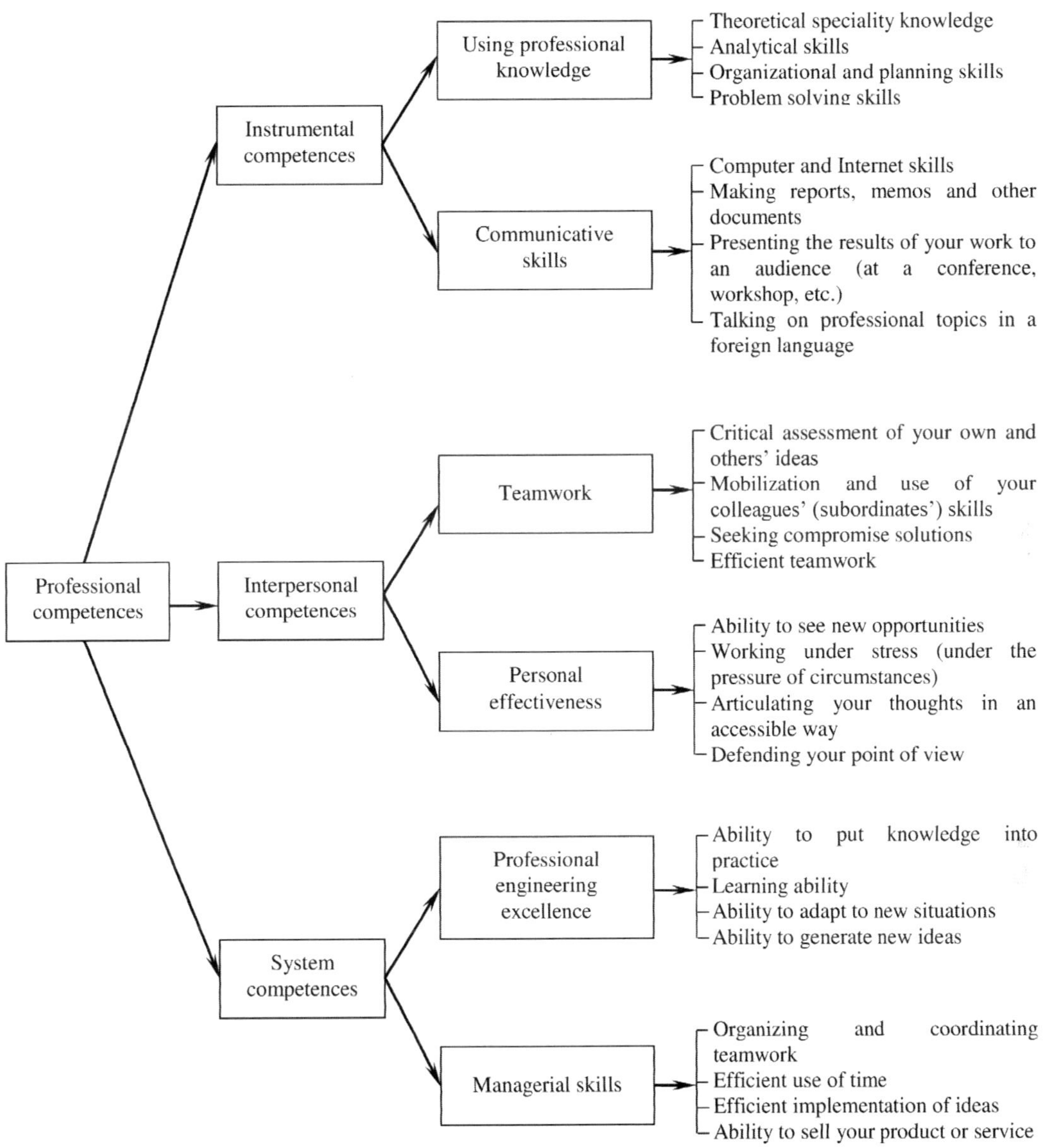

Figure 1. Classification of the engineering personnel competences.

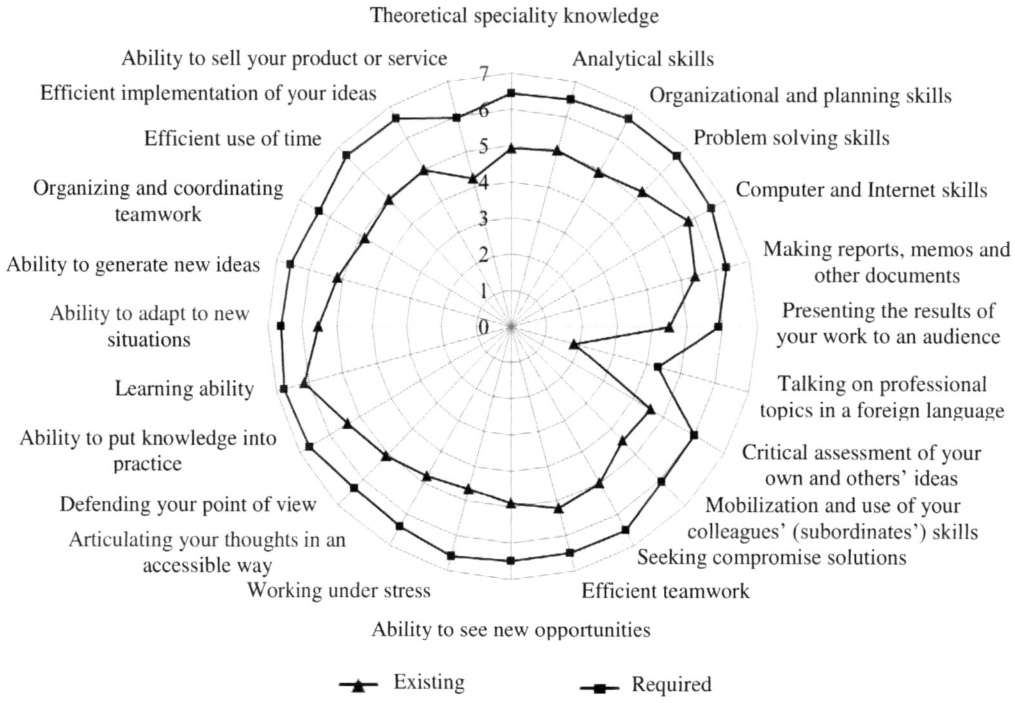

Figure 2. Competence-based profile of highly-qualified personnel.

4. Conclusions

The conducted research makes it possible to state that under present-day conditions higher school students should develop their organization and self-organization skills, effective communication skills, ability to perceive the new and learning ability in order to be successful and in-demand. Possessing these skills will make it possible to close the gap between the existing and the required competences. Considering that such competences as professional agility, results orientation and commitment to teamwork are being increasingly demanded, having just technical knowledge and skills to solve highly specialized engineering problems is no longer enough for the current labour market of the iron and steel industry. It is required to actively involve engineering personnel in innovation processes, such as development and adoption of advanced technologies for making steel grades with totally new or significantly enhanced properties.

References

[1] Zimnyaya I A 2013 *Writings of the National Association for Applied Linguistics* **4** 16–31
[2] McClelland D C and Dailey C 1974 *Professional Competencies of Human Service Workers* (Boston: McBer)
[3] Boyatzis R 1982 *Engineering The competent manager: A model for effective performance* (New York: John Wiley & Sons)
[4] Shmatko N A 2012 *Foresight* **4** 32–47
[5] Verbitsky A A and Larionova O L 2009 *Competence-based and person-centered approaches in education. Problems of integration* (Moscow: Logos)
[6] Spencer L M and Spencer S M 2005 *Competence at work. Translated from English* (Moscow: HIPPO)
[7] Parry S B 1996 *Training* **33** 48–56
[8] Lueneburger C, McClenaghan S, Paret F et al. 2012 *Competency and Potential: A Model for Private Equity Operating Partners* (New York: Egon Zehnder International)

[9] Whiddett S and Holliforde S 2008 *The Competencies Handbook. Translated from English* (Moscow: HIPPO)
[10] Solomennikova S I and Yakimovich B A 2012 *Bulletin of Izhevsk State Technical University* **3** 136–139
[11] Wilson R A, Cörvers F, Gardiner B et al. 2010 *Skills supply and demand in Europe. Medium-term forecast up to 2020* (Luxembourg: Publications Office)
[12] Wilson R A and Livanos I 2010 *Approaches to Modelling the supply of skills. Cedefop project on forecasting skill supply and demand in Europe. Technical Report no. 003* (Luxembourg: Publications Office)
[13] Lievens F, Sanchez J I and de Corte W 2004 *Personnel Psychology* **57** 881–904
[14] Allen J and van der Velden R 2011 *Flexible Professional in the Knowledge Society: New Challenges for Higher Education* (Dordrecht Heidelberg London New York: Springer)

Acknowledgements. This research was financially supported by the Ministry of Education and Science of Russian Federation within the framework of the project part of state task of Cherepovets State University (№ 11.3943.2017/ПЧ).

Intelligent Support System of Steel Technical Preparation in an Arc Furnace: Functional Scheme of Interactive Builder of the Multi Objective Optimization Problem

O S Logunova[1] and N S Sibileva[1]

[1] Nosov Magnitogorsk State Technical University, 38 Lenin street, Magnitogorsk, 455000, Russia

E-mail: logunova66@mail.ru

Abstract: The purpose of the study is to increase the efficiency of the steelmaking process in large capacity arc furnace on the basis of implementation a new decision-making system about the composition of charge materials. The authors proposed an interactive builder for the formation of the optimization problem, taking into account the requirements of the customer, normative documents and stocks of charge materials in the warehouse. To implement the interactive builder, the sets of deterministic and stochastic model components are developed, as well as a list of preferences of criteria and constraints.

1. Introduction

Modern metallurgical production is highly technological and science-intensive, both in terms of technology and management. In spite of this, there are a processes, where intellectual support tools are not used to make a decisions [1–8]. Decision-making is carried out on the basis of expert assessments, the significance of which is determined by the human experience. One such process is the decision-making about the using of stock of charge materials for the technical preparation of steel production in an arc furnace. When building the system, there is a need for the organization and maintenance of specialized software for an automated system of technical preparation of production, including the formation of a database with a ready-made solutions [9]. Compliance with the requirements for the chemical composition of steel, the cost of finished products, the saving of energy resources leads to the formulation of disparate tasks of multi-criteria optimization that do not have a relationship between the obtained solutions [10–13]. That is why, there is a need to create a system of continuous monitoring and process management [5], which requires automation of management processes and decision-making in the automated mode at all stages of metallurgical operations. In accordance with the above, the authors identified one of the problem in the synthesis of special mathematical software and a package of standard modules for the functioning of the intellectual support system for the technical preparation of steel production in an arc furnace.

2. Simulation

An analysis of the methods for preparation a technological letter on the structure of charge materials made it possible to develop new requirements to a system for collecting and preparing information.

The fundamental principles of the new system are: empiricalness, apriorism, adaptability, intellectuality.

The structure of the new system for collecting and preparing information for forming decisions about the structure of charge materials includes modules: interactive selection of attributes from the data warehouse; transformation and data generation; formation of a multicriteria optimization problem; intellectual decision-making support and the accumulation of a solutions for their subsequent adaptation and use [9]. The disadvantages of the previously known mathematical models are: the model restriction by the steelmaking conditions for one type of units DSP-180; the limited number of constituents of charge materials with the main types of charge materials (scrap metal, liquid and solid cast iron, hot briquetted iron); no restrictions on the reserves of charge materials.

To eliminate these shortcomings, an extended deterministic and statistical model is proposed to predict the results of arc melting.

The mathematical model includes two components: deterministic and statistical.

The set of deterministic components of model D includes known dependencies, assumptions, conditions and constraints. Such components are:

D_1: dependence $R = f(t_1, t_2)$, where R is specific electricity consumption per ton of finished steel, kW; t_1 is arc time under the current, minutes; t_2 is duration of melting, minutes;

D_2: system of constraints $[E_i^{\min}] \le [E_i] \le [E_i^{\max}]$, $i = \overline{1; k}$, where k is the number of chemical elements requiring strict limitations on their content for obtaining the designated steel grade; $[E_i]$ is proportion of the i-th chemical element in finished steel, %; $[E_i^{\min}]$, $[E_i^{\max}]$ is allowable proportion of the i-th chemical element in the finished steel according to the technological handbook, %;

D_3: system of constraints $[P_j^{\min}] \le [P_j] \le [P_j^{\max}]$, $j = \overline{1; m}$, where m is the number of components of the charge materials, %; $[P_i]$ is proportion of the j-th component of the charge materials in electric arc furnace, %; $[P_j^{\min}]$ and $[P_j^{\max}]$ is allowable proportion of the j-th component of the charge materials in electric arc furnace according to normative documents, %;

D_4: dependence $R_0 = f(\{[P_j]\})$, $j = \overline{1; m}$, where R_0 is bulk density of charge materials in the working area of an electric arc furnace, kg/m^3;

D_5: dependence $C = f(\{c_j\}, \{[P_j]\}, R(t_1, t_2), c_e)$, $j = \overline{1; m}$, where C is prime cost of one ton of finished steel, RUB, in thousands; c_e is cost of 1 kW of electric energy consumed by EAF, RUB;

D_6: system of constraints $0 \le [M_j] \le [M_j^{\max}]$, $j = \overline{1; m}$, where $[M_j]$ is mass of the j-th component of the charge materials, t; $[M_j^{\max}]$ is mass of j-th component of charge materials in production reserves, t.

The set of stochastic components of the model S assumes a description of the regularities for which analytical forms of recording based on the fundamental laws of physics, chemistry, and mathematics have not been known so far.

It is impossible to study such processes on operating aggregates and it is connected with the absence of methods for direct measurement of values, studied quantities, in real time, the presence of many known and unknown factors that influence the course of the process, the presence of high temperatures and enclosed space that restricts access to the object. The set of stochastic components of the mathematical model contains empirical dependences of the linear and nonlinear forms, which are constructed from the results of passive observation of the smelting process and are recorded in the melting report. The set of stochastic components includes empirical dependencies (1)–(5):

$$S_1 : t_1 = f(\{[P_j]\}), j = \overline{1; m}; \quad (1)$$

$$S_2 : t_2 = f(\{[P_j]\}), j = \overline{1; m}; \quad (2)$$

IOP Conf. Series: Materials Science and Engineering **287** (2017) 012009 doi:10.1088/1757-899X/287/1/012009

$$S_3 : [E_i] = f(\{[P_j]\}), j = \overline{1;m}; i = \overline{1;k}; \tag{3}$$

$$S_4 : [E_i] = f(\{[P_j]\}), j = \overline{1;m}; i = \overline{1;k}; \tag{4}$$

$$S_5 : T_s = f(\{[P_j]\}), j = \overline{1;m}. \tag{5}$$

The sets of deterministic and stochastic components expand depending on the requirements of the customer for the quality of the finished product, taking into account the technological conditions of the subsequent redistribution.

3. Result and analysis

The model forms the basis of the intellectual support system in the system of management of the technical preparation of production for the electric arc furnace. Paper [14] propose one of the classification of the of decision-making tasks, according to the available information on the variety of alternatives *X* and the optimality principle *opt*, the tasks are divided as follows:

– common decision-making problem – *X* and *opt* are unknown;
– the choice problem – *X* is known, *opt* is unknown;
– the problem of optimization or the problem of ordering the alternatives – *X* and *opt* are known.

The presence of a deterministic and stochastic model makes it possible to determine the structure of an interactive biulder for setting the problem of making decisions in the system for managing the technical preparation of production for an electric arc furnace, in particular, the choice of the structure of charge materials. In Figure 1 the structure of the interactive constructor is shown.

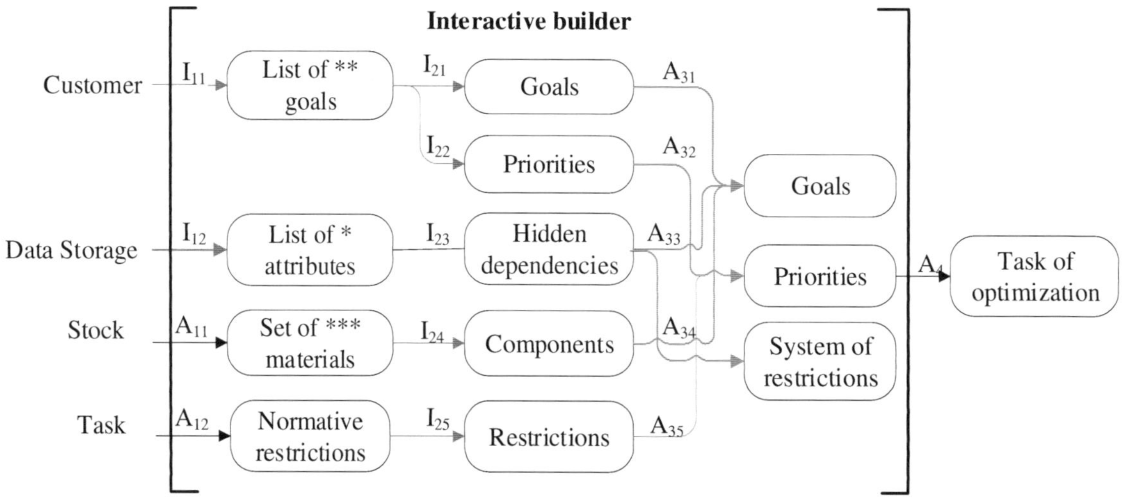

Figure 1. Scheme of functioning of the interactive builder of the multicriteria optimization problem for selecting components for steelmaking in EAF

The notations in Figure 1 are: I_{kl} – interactive mechanism for transferring data from sets *S* and *D*; A_{kl} – automatic data transfer mechanism from the sets *S* and *D*. Distinctive features of the interactive builder are: * – consolidation of empirical, normative and deterministic information in the formation of the model; ** – the possibility of forming a multiple goal; *** – the possibility of integration with the tasks of planning stocks in production warehouses.

4. Conclusion

Thus, the authors carried out:

1) rationale the need to set the task of multicriteria optimization in an interactive mode in the intellectual decision support system for the process of technical preparation of steel production in an arc furnace;

2) the construction of mathematical support in the form of a deterministic and stochastic model for its use in the module for the formation of a multicriteria optimization problem;

3) the development of a functional design of the interactive designer for the task of multi-criteria optimization for a new system of intellectual support for the technological preparation of steel production in an arc furnace.

References

[1] Bigeev V A, Valiahmetov A H , Burak J and Fedyanin A N 2014 *Vestnik of Nosov Magnitogorsk State Technical University* **1** 15–8
[2] Borshchov S M, Prohorenko V N and Vostrikov S S 2012 *Foundry and metallurgy* **2(65)** 63–5
[3] Zyuban N A, Gonik I L and Novickij N A 2014 Technological properties of the briquetted charge material from iron-containing wastes *Izvestia VSTU* **9** 138–40
[4] Kuznecov M S, YAkushev E V, Kulagin S A, Kotel'nikov G I, Semin A E and Chengeliya R K 2011 *Elektrometallurgiya* **1** 16–21
[5] Logunova O S, Matsko I I and Posochov I A 2013 *Vestnik of Nosov Magnitogorsk State Technical University* **5** 50–5
[6] Steblov A B 2016 *Foundry and metallurgy* **1** 66–1
[7] Tyutyukov S A and Andreev A V 2013 *Agrarnoe obrazovanie i nauka* **4** 1
[8] Hassan A I, Kotel'nikov G I, Semin A E and Megahed G 2015 *Chernye Metally (Ferrous metals)* **5(1001)** 64–9
[9] Logunova O S, Sibileva N S and Pavlov V V 2016 *Steel in Translation* **46** 733–38
[10] Logunova O S, Filippov E G, Pavlov I V and Pavlov V V 2013 *Steel in Translation* vol **43** 34–8
[11] Logunova O S and Pavlov V V 2014 *Metallurgist* **58** 299–5
[12] Logunova O S and Sibileva N S 2015 *Proc. Inf. Conf. on Computer Science and Information Engineering* (Bangkok: DEStech Publications) pp 394–9
[13] Nikiforov B A, Bigeev V A, Panteleev A V and Usherov A I 2007 *Vestnik of Nosov Magnitogorsk State Technical University* **1** 38–1
[14] Rykov A S 1999 *Methods of system analysis: multi-criteria and fuzzy optimization, modeling and expert evaluations* (Moscow: Publishing House "MISiS")

IOP Conf. Series: Materials Science and Engineering **287** (2017) 012010 doi:10.1088/1757-899X/287/1/012010

Study of the Impact of Heat Treatment Modes on Formation of Microstructure and a Given Set of Mechanical Properties of High-Strength Flat Products with Guaranteed Hardness (400 to 450 HB) from Low-Alloyed Steel

M Yu Matrosov[1], P G Martynov[1], T V Goroshko[1], M I Zvereva[1], A V Mitrofanov[2], K Yu Barabash[2]

[1] Federal State Unitary Enterprise I.P. Bardin Central Research Institute for Ferrous Metallurgy, 23/9 Radio Street, Bld. 2, Moscow, RU-105055, Russia
[2] PAO Severstal, 30 Mira Street, Cherepovets, RU-162608, Russia

E-mail: pscenterchermet@gmail.com

Abstract. The results of the study of influence of heat treatment modes on microstructure, size and shape of grains, mechanical properties of high-strength flat products from low-alloyed C-Mn-Cr-Si-Mo steel microalloyed by boron are presented. Heat treatment modes, which provide a combination of high impact viscosity at negative temperatures and guaranteed hardness, are determined.

1. Introduction

Production of high-strength rolled products (with hardness of 400 HB and more) with high impact viscosity, especially at negative temperatures, in combination with required mechanical properties is a challenging issue for steels used for manufacturing of mining equipment parts operated under conditions of high wear and impact stress.

Currently applied high-strength wear resistant steels represent low-alloyed C-Mn-Si steels containing about 0.15 to 0.25% of C, 0.8 to 1.5% of Mn, up to 0.6% of Si, usually micro-alloyed by boron and with the microstructure of lath high-dislocation martensite [1–3]. Hardness in such steels is proportional to carbon content in martensite [4, 5]. Mo, Cr, Ni and other elements are often added into steel to increase hardenability and resistance to softening during tempering, however, the amount of such elements is limited by requirements to weldability [2, 5, 6]. Increase of carbon content in martensite results in decrease of impact viscosity and cold resistance [4]. It is possible to increase impact viscosity of high-strength wear resistant steels by tempering, however, tempering at high temperatures causes considerable softening of steel, and low temperature tempering does not lead to significant improvement of the impact viscosity value [7, 8]. Moreover, reduction of impact energy because of cementite precipitation on boundaries of martensite laths is observed at low-temperature tempering in the temperature range of 300 to 400°C. It is necessary to find additional ways to increase impact viscosity. Decrease of structure elements – grains or martensite packets – is an effective way to enhance strength and brittle fracture resistance at the same time. It is known that the martensite packet size directly depends on the initial austenite grain size. Decrease of the martensite packet size by means of reduction of the austenite grain size before quenching results in significant increase of steel viscosity characteristics [15].

Heat treatment, specifically quenching and tempering, provides a given set of mechanical properties in finished rolled products thanks to changes in the size and morphology of structural components [16]. Correct selection of heat treatment modes contributes to generation of austenite fine-grained structure before quenching and generation of the target structure providing a required set of steel properties. Addition of microallying elements also contributes to that [17].

The goal of this paper is to study the influence of heat treatment temperature parameters on formation of the microstructure, which ensures a combination of high strength and viscosity characteristics and hardness in flat products.

2. Materials and Research Methods

Samples of rolled products of industrial production out of low-alloyed C-Mn-Cr-Si-Mo steel microalloyed by boron were used for conducting the study. Chemical composition of the steel being studied is presented in Table 1. The study of kinetics of phase transformations with construction of a continuous cooling transformation diagram of austenite transformation at continuous cooling, provided on Figure 1, was carried out for steel of given composition. It can be seen that the martensite microstructure is formed in case of cooling at speed more than 20°C/sec. Critical points of phase transformations at steel heating, determined experimentally, represent the following values: A_{c1}=740°C and A_{c3}=840°C.

Table 1. Chemical composition of study steel.

Elements mass fraction, %						CEV,	CET,
S	Si	Mn	P	S	Cr,Ni, Mo, Cu, Al, V, Ti, B, Nb, etc.	%	%
0.24	0.61	1.01	0.009	0.002	≤ 2.0	0.57	0.40

Note: Calculation of values of CEV and CET carbon equivalents is performed according to formulas:
CEV=C+Mn/6+(Cr+Mo+V)/5+(Cu+Ni)/15, %;
CET=C+(Mn+Mo)/10+(Cr+Cu)/20+Ni/40, %,
where C, Mn, Cu, Cr, Ni, Mo, V are mass fractions of chemical composition elements.

Flat products were manufactured by hot rolling on Mill-5000 at PAO Severstal. The study of influence of heat treatment modes on structure and mechanical properties was conducted on samples, thermally treated in laboratory furnaces. Temperature of heating of flat samples during quenching ranged from 770°C to 950°C with an interval of 30°C. After quenching, samples were subject to low temperature tempering. The received specimens were used for making samples for tensile testing (GOST 1497-84) and impact bending testing (GOST 9454-78), and for measuring hardness and studying microstructure, including through the use of transmission electron microscopy means. In order to determine boundaries of the initial austenite grain, pickling was performed in boiling oversaturated water solution of picric acid with addition of surface active agents. Quantitative parameters of microstructure were evaluated by Image Expert Pro 3 program.

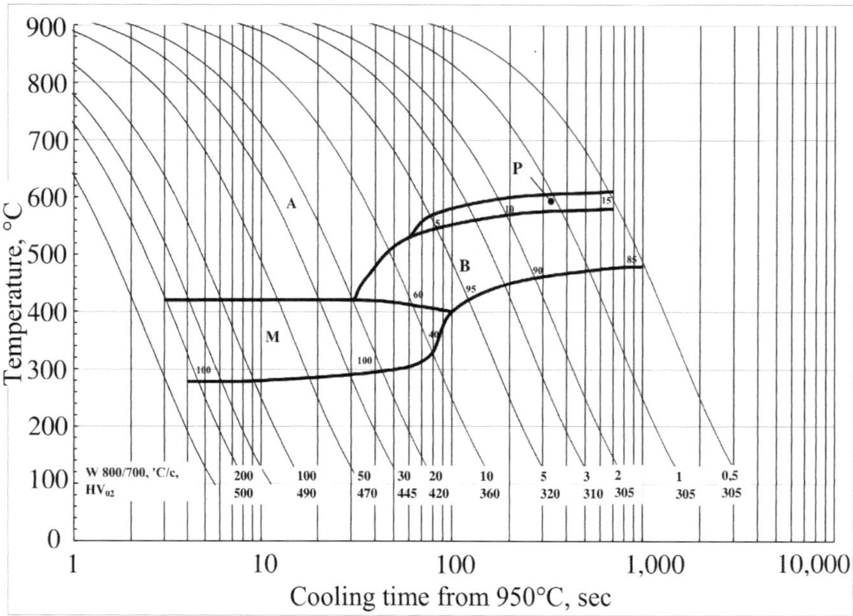

Figure 1. Continuous cooling transformation diagram of high-strength steel after heating up to 950ºC and cooling in the range of speed of 0.5 to 200ºC/sec.

3. Study Results and Their Discussion

Figure 2 shows a typical view of microstructure at different heat temperatures for quenching after special pickling to determine boundaries of prior austenite grains.

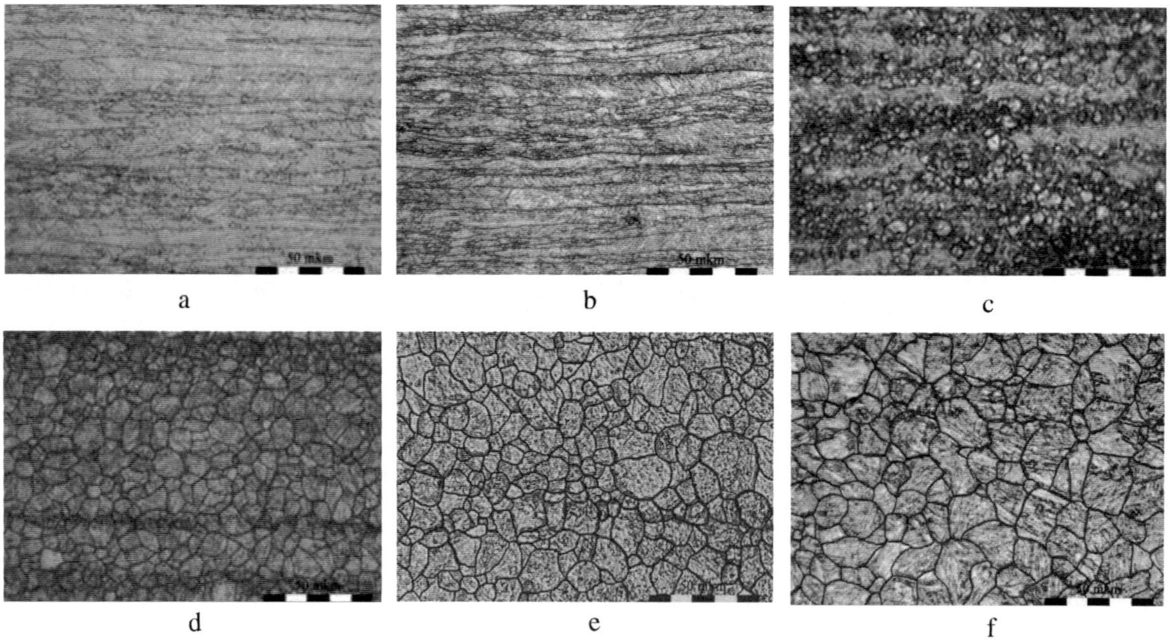

Figure 2. View of steel microstructure at different heat temperatures for quenching, x 400 (a) 770°C; (b) 800°C; (c) 860°C; (d) 890°C; (e) 920°C; (f) 950°C.

The microstructure analysis demonstrates that the shape of grains of steel being studied is changing from elongated to equiaxed with increase of heating temperature, which is related to the process of α→γ-transformation at heating. Grains have elongated shape (Figure 2 (a), (b)) at heating temperature for quenching in the range of 770 to 830°C, and new equiaxed grains (Figure 2 (c) start to appear at further increase of heating temperature at the boundaries of elongated grains, that is, the process of initial recrystallization takes place ending at the temperature of about 890°C (Figure 2 (d)). The process of accumulative recrystallization is observed upon further increase of temperature of heating for quenching: grains are growing due to integration with other adjacent grains (Figure 2 (e), (f)).

The average diameter of equiaxed grains depending on the temperature of heating for quenching was evaluated by Image Expert Pro 3 program. The average grain size grows from 8.5 µm to 14 µm with increase of temperature of heating for quenching from 890°C to 950°C, which corresponds to grain number 9 – 10 according to GOST 5639-82, and the grain maximal diameter at these temperatures ranges from 17.5 to 40.5 µm, which is shown on Figure 3.

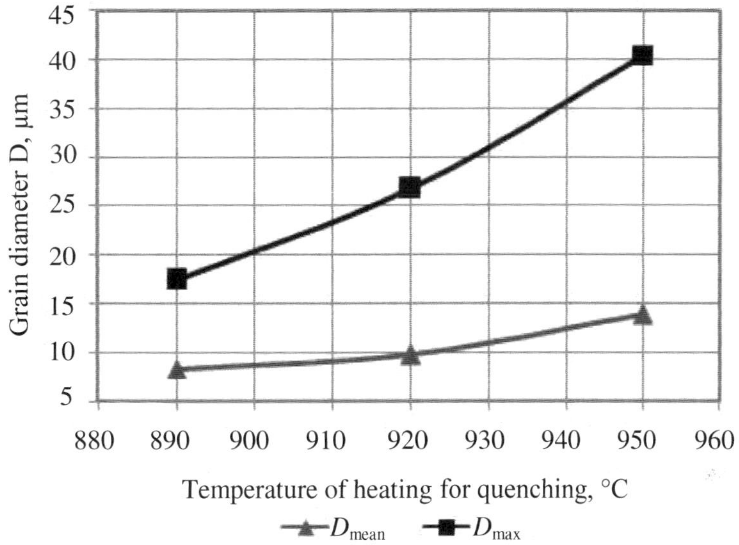

Figure 3. Diameter of the steel equiaxed grain depending on temperature.

Quenching from temperature of 890°C results in formation of the completely recrystallized structure consisting of quite small packets of tempered martensite of equiaxed shape (Figure 4 (a)). At the same time, high level of strength and impact viscosity is provided after low temperature tempering. The results of transmission electron microscopy (TEM) of a sample with quenching temperature of 890°C demonstrated that the structure of matrix consists of packets of high-dislocation lath martensite (Figure 4 (b)). The size of packets is from several microns to more than 10 µm, the typical width of laths is ~ 0.1 to 0.5 µm (individual laths up to ~ 1.5 µm). Plate-shaped carbide precipitates (cementite) were systematically identified inside martensite laths (Figure 4 (c)). Length of carbide plates is from few nanometres to ~ 150,200 nm. Precipitation of cementite films deteriorating impact viscosity were not observed on the boundaries of packets.

Figure 5 presents the influence of temperature of heating for quenching on mechanical properties of high-strength rolled products from low-alloyed C-Mn-Cr-Si-Mo steel microalloyed by boron.

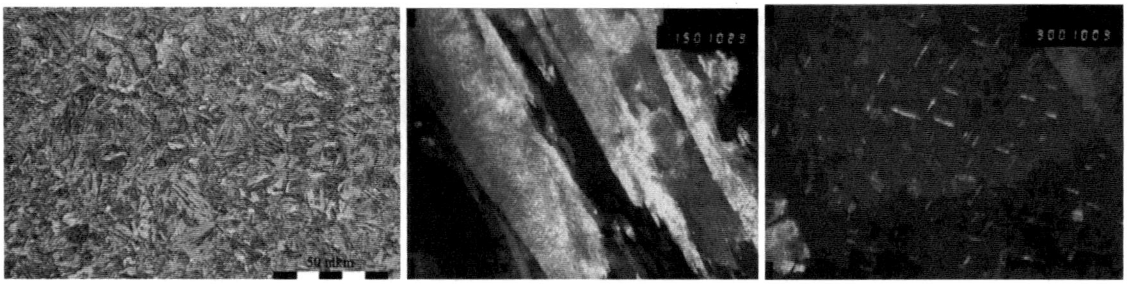

a b c

Figure 4. Microstructure of rolled strips from study steel after quenching with heating to 890°C and tempering: (a) optical microstructure (pickling in 4% solution of nitric acid (HNO_3) in ethyl alcohol), ×400; (b) lath martensite, TEM, ×15,000, dark field in martensite reflex; (c) cementite precipitation inside martensite laths , TEM, x 30,000, dark field in cementite reflex.

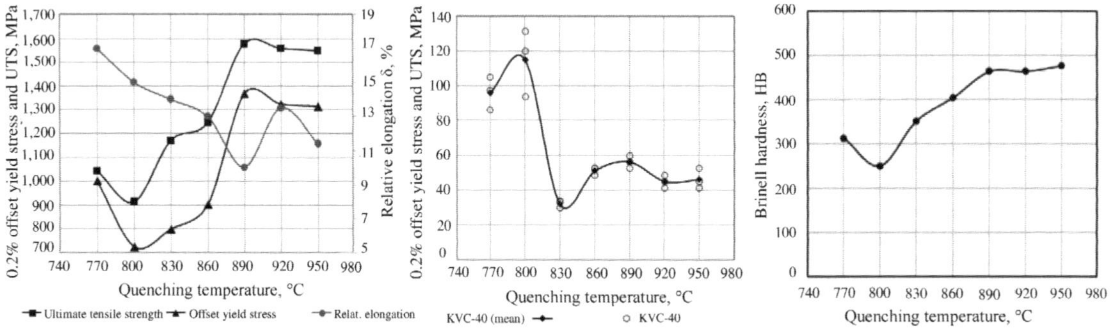

a b c

Figure 5. Influence of quenching temperature on mechanical properties of high-strength rolled products from study steel: (a) yield stress, ultimate tensile strength, relative elongation; (b) impact viscosity at temperature of -40°C; (c) hardness.

As one can see on Figure 5 (a), the highest values of strength performance – offset yield stress (1,365 MPa) and ultimate tensile strength (UTS) (1,575 MPa) – are achieved after heating up to 890°C, quenching and further tempering. Impact viscosity achieves the maximal value at temperature of -40°C after quenching from 800°C, and impact viscosity decreases first at quenching from higher temperatures, and then it gradually increases up to the value of 60 J/cm^2, which complies with requirements to steel used at negative temperatures under conditions of impact loading (Figure 5 (b)). Hardness value is over 400 HB when temperature of heating for quenching increases from 860°C (Figure 5 (c)). Quenching of rolled strips with higher heating temperature (950°C) results in modest reduction of strength properties and impact viscosity.

4. Conclusions

Variations of optimal modes of heat treatment providing a combination of high level of impact viscosity, strength properties and hardness of flat products are determined.

The obtained study results enable optimisation of processes of heat treatment of high-strength flat products from low-alloyed C-Mn-Cr-Si-Mo steel microalloyed by boron for certain application conditions.

IOP Conf. Series: Materials Science and Engineering **287** (2017) 012010 doi:10.1088/1757-899X/287/1/012010

References

[1] Mohrbacher H 2011 *Int. Symp. on the Recent Development in Plate Steels* (Colorado: Winter Park) pp 169–178
[2] Stalheim D G, Peimao F, Linhao G and Yongqing Z 2015 *HSLA Steels, Microalloying & Offshore Engineering Steels* pp 996–1001
[3] Deng X, Wang Z, Yan G and Wang G 2010 *Journal of Northeastern University* **31(7)** 942–46
[4] Irvin K J, Pickering F B and Carstone J 1960 *J. Iron Steel Inst.* **196** 66
[5] Grange R A, Hribral C R and Porter L F 1977 *Metallurgical transactions A* vol 8A pp1775–85
[6] Murota Y, Abe T and Hashimoto M 2005 *JFE Technical Report* pp 60-65
[7] Gojic M, Kosec L, and Matkovic P 1998 *Journal of Materials Science* **33(2)** 395–403
[8] Morito S, Saito H, Ogawa T and Furuhara T Maki T 2005. *J. ISIJ Int.* **45** 91-94
[9] Lement B S et al 1954 *Trans A.S.M.* **46** 851
[10] Klinger L.J. et al 1954 *Trans A.S.M.* **46** 1557
[11] Krauss G 2005 *Steels: processing, structure and performance* (USA: ACM International) p 343–346
[12] Zayats L Ts, Panov D O, Simonov Yu N, Titova M G, Sharlaimova E N and Balakhin A N 2011 *Bulletin of Perm National Research Polytechnic University. Mechanical Engineering, Material Engineering* **2** 54–61
[13] Smsirnov M A Schastlivtsev V M, Zhuravlev L G 1999 Basic Concepts of Steel Heat Treatment: Training (Yekaterinburg: Uro RAN) p 496
[14] Zhen F, Wang W, Yang H ,Chen S ,Zhang X and Qu J *Journal of Iron and Steel Research, Int.* **18** 637-640
[15] Wang C, Wang M, Shi J, Hui W and Dong H 2008 *Scripta Materialia* **58** 492–495
[16] Kotrechko S A, Meshkov Yu Ya and Televich R V 2006 *Metal Science and Heat Treatment.* **9** 28-34
[17] Bian J, Mohrbacher H, Zhang J, Zhao Y, Lu H and Dong H 2015 *Advances in Manufacturing* **3** 27–36

IOP Conf. Series: Materials Science and Engineering **287** (2017) 012011 doi:10.1088/1757-899X/287/1/012011

Experience of Application of Liquid Lubricating Materials during Wide Strip Hot Rolling

S I Platov[1], R R Dema[1], M V Kharchenko[1], R N Amirov[1]

[1] Federal State Budgetary Educational Institution of Higher Education G. I. Nosov Magnitogorsk State Technical University, 38 Lenina Prospect, Magnitogorsk, RU-455000, Russia

E-mail: ruslan246@mail.ru

Abstract. The paper presents the results of the scientific and practical research of roller systems operation at feed of liquid lubricating materials through the example of the wide strip hot rolling Mill-2000 at PAO MMK. The experiments proved that application of lubricating materials leads to decrease of energy-power parameters of the process by 12 to 15 %, and reduction of work roll wear by 10 to 12%. The practical results of the study are developed recommendations on determination of consumption-volumetric parameters of the supplied lubricating material depending on rheological and geometrical parameters of the rolled strip and current wear of work rolls.

Nowadays, PAO MMK performs considerable work on assimilation and implementation of new steel grades of high strength classes capable to withstand high loads and operation in rough environments, etc. During manufacture of such products, the production equipment, in particular hot rolling mills, is subject to high loads with respect to energy-power parameters close to critical values. One of practical ways aimed at reduction of energy-power parameters of the process is application of systems of liquid process lubricant supply (SLS) in hot rolling production technology.

Application of lubricating materials in hot rolling technology has been known for a long time [1–3]. It is known from the literature that application of lubricating materials during hot rolling leads to significant reduction of energy-power parameters of the process [1]. However, application of the technology of lubricant supply on to the rolls surface has not been widely used, since solid lubricants and graphite-based greases (rarely, liquid ones) were used as lubricants most frequently. Application of graphite-based lubricating materials results in inequal friction forces along the perimeter of the deformation zone, which in turn leads to deterioration of rolled strip shape [3]. Currently, the wide strip rolling technology with application of process lubricant has been given new impetus. First of all, it is related to assimilation of a new range of high-strength steels, and simultaneous development of new liquid lubricating materials with high adhesive properties capable to withstand high temperatures for short moments [4, 5].

In 2008, the system of liquid process lubricant supply to the first three stands (7, 8, 9) of the continuous finishing train of the rolling mill was installed on the continuous wide strip hot rolling Mill 2000 at PAO MMK. A special feature of the installed SLS is that the lubricating material (LM) is supplied as prepared water and oil dispersion to surfaces of work rolls at the side of metal exit from the stand. The machine overview diagram is provided on Figure 1.

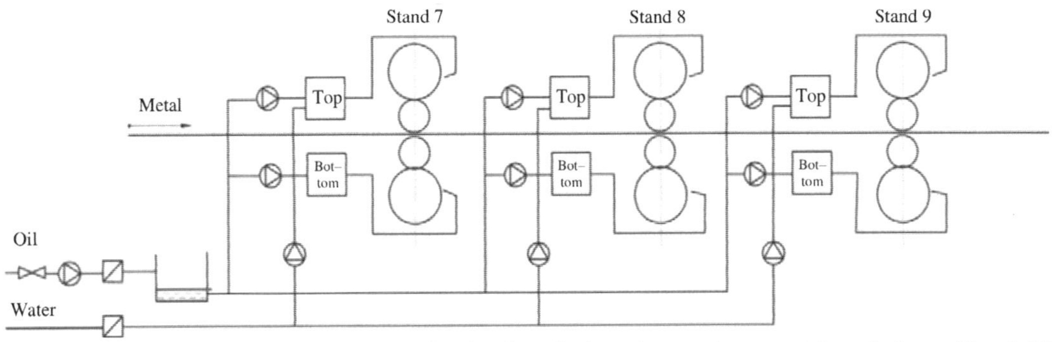

Figure 1. The system of process lubrication installed on the continuous wide strip hot rolling Mill 2000 at PAO Magnitogorsk Iron and Steel Works.

Practical application of the SLS demonstrated decrease of energy-power parameters and increase of service life of mill rolls during rolling [6–8].

Figure 2 presents comparative graphs describing the decrease of energy-power parameters of the rolling process with application of the SLS in average by 3 to 6%.

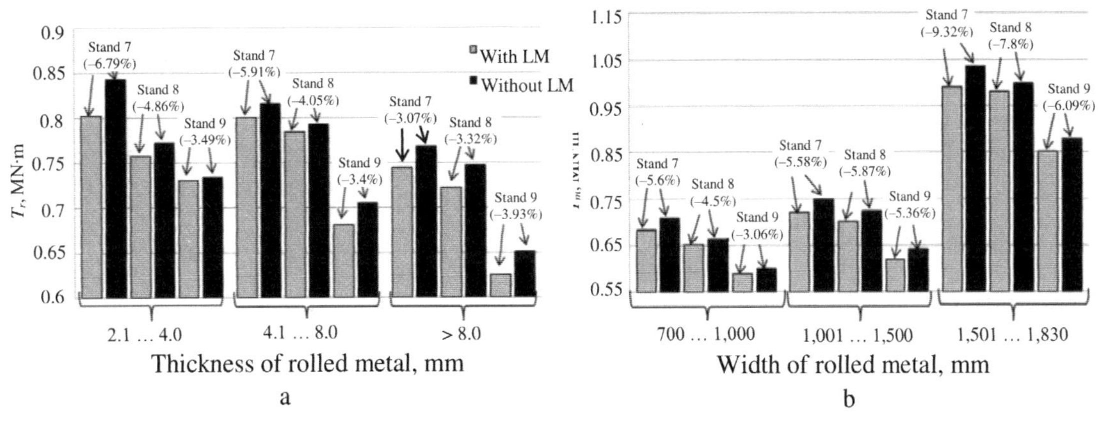

Figure 2. Averaged value of torque of the motor (T_m) depending on thickness (a) and width (b) of rolled steel for the 7th, 8th and 9th stands of the finishing train of the continuous wide strip hot rolling Mill 2000 at PAO MMK.

One of major deficiencies of the SLS is lack of a mathematical tool describing the correlation between the amount of the LM supplied and properties of rolled material; lack of recommendations and clear procedures for determination of process modes of LM supply. Thus, during first experimental adjustments of this system it was found that slip of work rolls against the rolled strip is possible at excessive LM supply, and in case of insufficient supply, the effect stated by a manufacturer of this unit is not observed [9]. In the early stage, the installed SLS was operated in the limited mode, and reduction in energy consumption was insignificant and averaged to 3 to 6% [4, 5].

Since 2008, the team of Federal State Budgetary Educational Institution of Higher Education G. I. Nosov Magnitogorsk State Technical University has been actively studying the process of wide strip hot rolling with application of the lubrication system. Based on the results of conducted studies, practical recommendations on adjustment of the SLS were received:

1. An adequate mathematical dependence, which determines consumption-volumetric parameters of the supplied LM depending on rolling process parameters, was received [15]:

$$Q = Ln_{br}\left[\left(\left(\frac{\beta}{180\pi}+\left(90-\frac{2l_c}{D}\right)\right)K_r\Sigma R_z\right)+\left(\left(2.256\cdot\sqrt{\eta\cdot q_m\cdot\frac{R_1\cdot R_2}{R_1+R_2}}\right)\left(\frac{3.17\eta_0^{0.75}\alpha^{0.6}\upsilon_{rl}^{0.75}}{q_m^{0.15}\rho_r^{0.4}}\right)\right)+\left(\left(\frac{3\eta_0\cdot\alpha\cdot\upsilon_{rl}}{\alpha_{cap}\cdot[1-e^{-\alpha\sigma_T^*}]}+k_r\Sigma R_z\right)\left(\frac{\omega_b+\omega_w}{\omega_w+\omega_{av}}\right)\right)\right]-\left(Q_a-\left(\frac{h_a-h_1}{h_a}\right)\cdot Q_a\right), \quad (1)$$

where β is an angle of LM application to the horizontal axis of the backup roll (degrees); R_1, R_2 are the radii of the backup roll and work roll respectively, mm; l_c *is* half-width of the contact area, mm; D is backup roll diameter, mm; n_{br} is the number of revolutions of the backup roll, 1/min; q_m is load per unit of length (metal pressure onto the roll), N/m; η_0 is oil dynamic viscosity at atmospheric pressure and operating temperature, Pa·s; υ_{rl} is rolling speed, m/s; α is pressure-viscosity coefficient; ρ_r is relative curvature of rolls surfaces, m; ω_b is backup roll angular speed, s^{-1}; ω_w is work roll angular speed, s^{-1}; ω_{av} is average speed of backup and work rolls, s^{-1}; η_0 is LM dynamic viscosity, Pa·s; θ is pressure-viscosity coefficient, $Pa^{-1}\cdot s^{-1}$; α_{cap} is LM capture angle, rad.; σ_T^* is yield strength of rolled billet considering hydrodynamic effect in the near-contact zone, $\sigma_T^* = 1.15\sigma_T$, MPa; Q_a are losses of the lubricating material caused by adhesion with surface of the roll; k_r is the coefficient describing the intensity of LM grip by surface roughness, $k_r = 0.09 \div 0.015$; ΣR_z is total roughness, μm; h_1 is LM thickness at the exit from the interroll contact, mm; h_a is thickness of the LM adhesive layer.

2. The mathematical algorithm of SLS operation implemented numerically was developed and adopted considering technical special features. This mathematical algorithm performs automatic adjustment of the system depending on the type and size of the billet being rolled. Table 1 provides recommended modes of SLS operation.

Table 1. Recommended and permissible values of LM consumption per stand of the continuous wide strip hot rolling Mill 2000 at PAO MMK [15].

Classification of rolled products range based on deformation resistance parameter (σ_{0i})*	Recommended and [permissible] consumption per stand (l/min)		
	Stand 7	Stand 8	Stand 9
< 76 MPa	0.10 to [0.11]	0.15 to [0.16]	0.31 to [0.32]
76 – 82 MPa	0.14 to [0.16]	0.13 to [0.15]	0.21 to [0.24]
82 – 94 MPa	0.08 to [0.10]	0.10 to [0.12]	0.15 to [0.17]
94 – 102 MPa	0.07 to [0.09]	0.11 to [0.12]	0.13 to [0.16]
102 – 116 MPa	0.06 to [0.08]	0.09 to [0.10]	0.14 to [0.19]

* deformation resistance of the range of rolled products (σ_{0i}) was determined according to the Andreyuk-Tyulenev method.

IOP Conf. Series: Materials Science and Engineering **287** (2017) 012011 doi:10.1088/1757-899X/287/1/012011

The implementation improved the system performance by 12 to 15% in average.

Further operation and study of the SLS demonstrated that application of the lubricating material at hot rolling leads to reduction of wear of work rolls. Based on the studies described in papers [10–13], the authors theoretically determined and justified reduction of energy consumption in the area of deformation during rolling with application of the lubricating material [14, 15]. In such a case, it was determined that wear reduction was observed not only in the first three stands, where the SLS was installed (Figure 3), but in all other stands of continuous finishing train of Mill 2000 at PAO MMK. The results of comparative measurement of the volumetric wear (weight loss) of work rolls with LM supply indicate that roll wear reduction averages 10 to 15%.

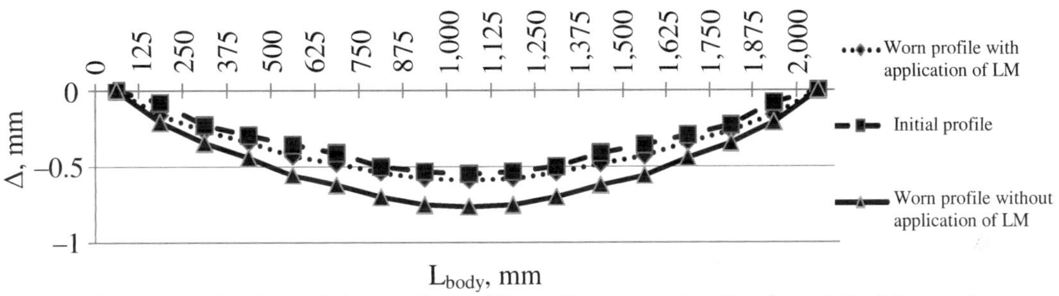

Figure 3. Averaged values of changes in profiling of upper work rolls of stand 7 of the continuous wide strip hot rolling Mill 2000 at PAO MMK.

In addition to wear reduction, it was determined that the SLS also has a positive impact on reduction of the amount of metal removed from the surface of work rolls during further regrinding by 3 to 5% in average.

Practical results of the research are:

– Development and implementation of the upgraded mathematical apparatus capable to perform adjustment and control of the SLS depending on the degree of wear of work rolls.

– Reduction of energy-power parameters during rolling by 10 to 18% in average depending on the range of rolled products.

– Reduction of work rolls wear by 10 to 12%.

References

[1] Belosevich V K, Netesov N P, Meleshko V I and Adamsky S D 1976 *Emulsion and Lubricants at Cold Rolling* (Moscow: Metallurgy) p 416

[2] Kokroft M G 1970 *Lubrication and Lubricating Materials* (Moscow: Metallurgy) p 11

[3] Tuboltsev L G 1982 *Sheet Steel Hot Rolling with Process Lubricants. College textbook* (Moscow: Metallurgy) p 160

[4] Dubrovsky S V, Dema S V, Kharchenko M V and Yaroslavtsev A V 2011 *Manufacturing of Rolled Products* **12** 6–8

[5] Plativ S I, Rumyantsev M I, Dema R R and Kharchenko M V 2011 Bulletin of G. I. Nosov Magnitogorsk State Technical University **4** 19–21

[6] Vorozhishchev A N, Dema R R and Kazakova T V 2016 *2nd Int. Conf. Proc. ICIE 2016* p 1007–1012

[7] Tyuteryakov N S, Dema R R and Nefed'ev S P 2016 *2nd Int. Conf. Proc. ICIE 2016* p 667-673

[8] Kharchenko M V, Dema R R and Bilichenko V I 2016 *Solid State Phenomena* vol 870 p 446–453

[9] Borovkov I V 2011 *Collected Papers of Central Laboratory Complex at OAO MMK No.15 Technology Improvement at OAO MMK* p 243–252
[10] Garber E A, Kozhevnikova I A and Tarasov P A 2007 *Manufacturing of Rolled Products* **4** 7–14
[11] Garber E A, Kozhevnikova I A and Tarasov P A 2007 *Manufacturing of Rolled Products* **10** 5–12
[12] Garber E A and Kozhevnikova I A 2010 *Manufacturing of Rolled Products* **12** 16–24
[13] Garber E A, Goncharsky A A and Petrov S V 2000 *Manufacturing of Rolled Products* **12** 9-13
[14] Dema R R, Kharchenko M V and Platov S I 2013 *The Theory and Practice of Process Lubrication Application during Wide Strip Hot Rolling* (Magnitogorsk: Publication of G. I. Nosov Magnitogorsk State Technical University) p 99
[15] Dema R R, Yaroslavtsev A V, Yaroslavtseva K K 2014 *Lubricating Material Supply to Rolls of Wide Strip Hot Rolling, and Evaluation of Process Energy Efficiency Using an Example of Mill 2000 at OAO MMK* (Magnitogorsk: Publication of G. I. Nosov Magnitogorsk State Technical University) p 72

Acknowledgements. The work was conducted with the financial support of The Ministry of Education and Science under the project No. 11.2054.2017/PCh (Rus. “11.2054.2017/ПЧ”) within the government assignment for 2017–2019 (project number 11.2054.2017/4.6).

IOP Conf. Series: Materials Science and Engineering **287** (2017) 012012 doi:10.1088/1757-899X/287/1/012012

Economical and Energy Efficiency of Iron and Steel Industry Reindustrialisation in Russia Based on Implementation of Breakthrough Energy-Saving Technologies

L N Shevelev[1]

[1] Federal State Unitary Enterprise I.P. Bardin Central Research Institute for Ferrous Metallurgy, 23/9 Radio Street, Bld. 2, Moscow, RU-105055, Russia

E-mail: economy@chermet.net

Abstract. Estimates were given of economical and energy efficiency of breakthrough energy-saving technologies, which increase competitive advantages and provide energy efficiency of production while reducing negative impact on the environment through reduction of emissions of harmful substances and greenhouse gases in the atmosphere. Among these technologies, preference is given to the following: pulverized coal fuel, blast-furnace gas recycling, gasification of non-coking coal in bubble-type gas-generators, iron-ore concentrate briquetting with steam coal with further use of ore-coal briquettes in electric furnace steel making. Implementation of these technologies at iron and steel works will significantly reduce the energy intensity of production through reduction of expensive coking coal consumption by means of their substitution by less expensive non-coking (steam) coal, and natural gas substitution by own secondary energy resource, which is the reducing gas. As the result, plants will get an opportunity to become self-sufficient in energy-resources and free themselves entirely from expensive purchased energy resources (natural gas, electric power, and partially coking coals), and cross over to low-carbon development.

The iron and steel industry in Russia is classified as the most energy-intensive industry. According to assessment of the Institute of Global Climate and Ecology of the Federal Service of Russia on Hydrometeorology and Monitoring of the Environment and Russian Academy of Science, the iron and steel industry accounts for about 60% of all emissions in the Industrial Processes sector reflected in the National Survey Cadastre of Greenhouse Gases Emissions in Russia.

22 projects of joint implementation were carried out in the iron and steel industry in Russia over the period of the Kyoto Protocol (2008 to 2012). Overall decline in greenhouse gases emissions in the industry in 2014 compared to 1990 amounted to 36.7% (64.8 mln tons of CO_2 equivalent). This is the largest reduction of emissions in the world iron and steel industry. It was achieved due to the following factors: decline in the production volume caused reduction by 58.4%; reorganisation of inefficient facilities caused reduction by 33.5%; improvements in energy efficiency of production caused reduction by 8.1%. In advanced industrial countries (the EU, the USA, Japan, etc.), indices of reduction of greenhouse gases emissions for this period equalled to 6–8%, which is at the level of the iron and steel industry in Russia due to improvements in energy efficiency of production. At the same time, Russia has the worst specific indicators on steel energy intensity (carbon footprint), which in 2014 were: the USA = 1.2 tonnes of CO_2 per tonne of steel; the EU = 1.4 tonnes of CO_2 per tonne of steel; Russia = 1.6 tonnes of CO_2 per tonne of steel (estimated by the author), which is in percentage terms respectively: USA/EU/Russia = 100.0% / 116.7% / 133.3%.

The problems of manufacturing of energy efficient and competitive steel products in Russia, and greenhouse gases emissions reduction in order to maintain climate and fulfil commitments undertaken under the Paris agreement, are the most pressing challenges of the twenty-first century. They require implementation of measures on maintenance of the Earth's surface temperature not greater than 2°C.

Inventory of commitments undertaken by the countries under the Paris agreement demonstrated that they do not ensure the accepted temperature criterion of 2°C, but will result in temperature growth to 3–4°C. In addition, according to the World Steel Association, steel output will grow by 2050 to 2.7 bln tonnes, which is by 70%. Emissions of greenhouse gases will increase in the same amount, i.e. by 70% in 2050 compared to 1990.

Energy-saving technologies should be rapidly implemented in order to avoid catastrophic climate change [1–3]. In advanced western countries, the priority has been given to the technology of pulverized coal fuel in the blast furnace process providing coke consumption at the level of 360–380 kg per tonne of hot metal [4, 5]. In Russia, the technology of iron making with application of natural gas has been developed considerably. It provides coke consumption at the level of 410–420 kg per tonne of hot metal. In Russia, the pulverized coal fuel technology is implemented at three enterprises: Novolipetsk Steel, Nizhniy Tagil Iron and Steel Works, West Siberian Metallurgical Plant, which provided the reduction of coke consumption to the level of 350 to 370 kg per tonne of hot metal.

Estimates of the State Scientific Centre of the Russian Federation Federal State Unitary Enterprise (FSUE) I.P. Bardin Central Research Institute for Ferrous Metallurgy together with the consortium of scientific and project organizations for use of the blast furnace gas recycling technology at new (domestic) equipment at PAO Tulachermet demonstrated the following results: coke consumption was reduced by 36 to 38%; natural gas was excluded from blast-furnace smelting by 100.0%; manufacturing costs in iron prime cost decreased by 15 to 20%; greenhouse gases emissions were reduced by 50 to 70%; blast furnace production capacity increased by 30 to 35% [7].

High energy intensity of steel products in Russia can become a serious obstacle for export. The EU and the USA are developing restrictive measures to prevent such products from entering their local markets (the standard on access of import steel with energy intensity indicators not higher than the indicators at the national level is being developed). The concept of energy-saving development of ferrous metallurgy till 2050 is under consideration in the USA. This concept involves liquidation of steel production in basic oxygen furnaces (BOF) together with the coke-sinter-blast furnace process of iron production and their substitution by electric furnace production based on use of direct-reduced iron and steel scrap. Thanks to implementation of this concept, steel energy intensity in the USA will equal to 0.8 tonnes of CO_2 per tonne of steel (estimated by the author). In Russia, the following technologies can be developed alternatively: recycling of blast-furnace gas + coal gasification in gas generators with complete withdrawal of natural gas and purchased electric energy at production of the BOF steel [8–11]; the ORIEN technology in the electric furnace production with complete refusal of consumption of purchased electric energy. It is possible that in future Russia will cross over to the electric furnace production based on the ORIEN technology.

The idea of the ORIEN technology is the following: a new charge material – ore-coal briquettes (it is possible to use steel scrap depending on demand of CO generation and quality of steel to be received) – is electrically melted under high temperature. A liquid-phase product – direct-reduced iron oversaturated with carbon – is received during blowing. Two products are received after oxygen blowing: steel and reducing gas CO. Thus, the electrical smelting “Ore – Steel – Energy” is becoming simultaneously blast-furnace and basic oxygen furnace conversions. The amount of generated CO is enough (about 1,100 m^3 per tonne of steel) to meet the demand of the steel making production for electrical energy in full. Together with coal generation in gas generators, it satisfies the need of the plant for own electrical energy completely in case of crossing over to the electric furnace production with use of ore-coal briquettes.

To move from the BOF steel production to electric furnace steel, an adaptation period is required for adoption of electric smelting with ore-coal briquettes with respect to the range of smelted steel grades and the parameters of raw materials. At some stage, both technologies will develop simultaneously (blast furnace gas recycling and ORIEN). The blast furnace – BOF technology has the highest potential of energy saving (BOF steel amounts to 60%, and EAF steel amounts to 40%). The paper [13] describes the study of economical and energy efficiency of energy saving technologies in blast furnace production compared to iron production without processing additives (natural gas, pulverized coal fuel, hot reducing gases and their combined technologies). The following results are received: when coke was used without energetic additives (500 kg/t of hot metal), its level with additives was: with natural gas 420 kg/t of hot metal; with pulverized coal fuel 340 kg/t; with hot reducing gas 290 kg/t; according to combined technology (pulverized coal fuel + hot reducing gases within the study of Mittal Steel Company [11]) 275 kg/t; according to the combined technology (hot reducing gases + gas generator) 270 kg/t. Changes in operation costs amounted to: 100%; 92.3%; 88.2%; 82.0%; 71.6%. Changes in emissions of greenhouse gases were: 100%; 87.4%; 84.6%; 53.4%; 59.4%; 36.4%. Thus, the best results by all accounts were received for the technology of hot reducing gases with a gas generator.

Similar studies were performed for electric furnace steel according to the ORIEN technology compared to BOF steel. In particular, operation costs for BOF steel produced according to the new technology (hot reducing gases with a gas generator) decreased compared to the basic technology by 25.5%.

In all options of electric furnace steel making, when ORIEN technology was used, operation costs decreased compared to the basic technology: for electric furnace steel making with direct-reduced iron by 18.2%; for electric furnace steel making with steel scrap by 6.9%. Similar dynamics of results was received after comparison of specific emissions (carbon footprint), which amounted to respectively: 54.2%; 61.5%; 61.6%; 4.3%.

Provided data demonstrate that application of the technology of hot reducing gases in blast furnace production and the ORIEN technology in electric furnace production leads to the best results with respect to economical and energy efficiency. Implementation of these technologies in the industry in general might save 36.0 bln m^3 of natural gas per year and 27.4 bln kWh of electrical energy per year. Total savings from implementation of these technologies in the industry in general will be 248.0 bln roubles per year with capital expenses of 280 bln roubles, i.e. the payback period is about one year. Besides, emissions of greenhouses gases will be reduced by 50%, which guarantees fulfilment of commitments of Russia undertaken according to the Paris agreement on climate.

References

[1] Shevelev L N and Brodov A A 2012 *Metallurgist* **8** 7–13
[2] Shevelev L N and Brodov A A 2012 *Electrometallurgy* **10** 2–10
[3] Soskovets O N, Shevelev L N and Brodov A A 2012 *Steel* **10** 81–87
[4] Smith T 2011 *Steel Times international* **35** 33-40
[5] Lelyuk V P et al 2014 *Steel* **1** 2–5
[6] Yusfin Yu S, Chernousov P I et al 2004 *Iron Metallurgy* (Moscow: Akademkniga Publishing Centre) p 774
[7] [6] Soskovets O N, Shevelev L N et al 2014 *Steel* **5** 103–107
[8] Ramm A N 1980 *Modern Blast Furnace Process* (Moscow: Metallurgy) p 304
[9] [3] Voskoboinikov V G, Gokhman Yu I and Zhurakovsky B L 1971 *Steel* 4 303–308
[10] [4] Buzoverya M T and Buzoverya V M 2002 *Steel* 2 13–15
[11] Jan Van der Stel, Gouverse G et al 2013 *Steel Times international* **31** 36–38
[12] Dorofeev G A, Yantovsky P R, Smirnov K G and Stepanov Ya M 2017 *Ferrous Metals* **5** 17–22
[13] Shevelev L N 2015 *Business, Organization, Strategy, Systems – BOSS* **11** 58–62

IOP Conf. Series: Materials Science and Engineering **287** (2017) 012013 doi:10.1088/1757-899X/287/1/012013

Scientific and Technological Principles of Development of New Cold-Resistant Arc-Steels (Steels for Arctic Applications)

O V Sych[1], E I Khlusova[1], E A Yashin[1]

[1] National Research Centre Kurchatov Institute – Central Research Institute of Structural Materials Prometey, 49 Shpalernaya str., St. Petersburg, RU-191015, Russia

E-mail: oknir@crism.ru

Abstract. The paper presents the results of quantitative analysis of C, Mn, Ni and Cu content on strength and cold-resistance of rolled plates. Relations between the ferritic-bainitic structure morphology and anisotropy and steel performance characteristics have been established. Influence of thermal and deformation rolling patterns on steel structure has been studied. The steel chemical composition has been improved and precision thermomechanical processing conditions for production of cold-resistant Arc-steel plates have been developed.

1. Introduction

Analysis of the current state of the shipbuilding industry shows that construction of high-technology marine facilities for Arctic applications (high duty ice-breakers, ice-going research vessels, off-shore ice-resistant platforms of various types, tankers, container ships, supply vessels, etc.) is the priority development area for the domestic shipbuilding. The Russian Maritime Register of Shipbuilding has developed requirements to steels for Arctic applications to be applied without limitation in the Arctic conditions for all types of structural elements. These requirements have been made part of "Rules ..." of the Russian Maritime Register of Shipbuilding and GOST R 2927-2015. Arc-index is given to steels which reliably meet the additional requirements at a desired temperature (T_d[1]). To ensure that there will be no brittle fracture, Arc-steels must meet the following additional requirements as to: ductile-to-brittle transition temperatures T_{cb}[2] and nil-ductility temperatures NDT[3], crack resistance characteristic (critical crack tip opening) CTOD (Crack Tip Opening Displacement) at T_d for base metal (BM) and heat-affected zone metal (HAZ) (Table 1) [1].

[1] T_d – minimum service temperature of material up to which a given steel grade may be used for all structural elements without limitations.

[2] T_{cb} – critical brittle point, at which during three-point ultimate static bending of a full-thickness specimen with a notch as a concentrator no less than 70% of the fibrous constituent is observed in the fracture.

[3] NDT – nil-ductility temperature defined as the maximum temperature at which destruction of a standard-size specimen with a brittle weld deposit and crack initiating notch takes pace under impact loading.

Table 1. Requirements to cold resistance and operational performance of Arc-steel grades F36W and F40W.

Rolled steel thickness, mm	NDT, °C			T_{cb}, °C			CTOD at T_d, mm			
							BM		HAZ	
	Arc30	Arc40	Arc50	Arc30	Arc40	Arc50	F36W^{Arc}	F40W^{Arc}	F36W^{Arc}	F40W^{Arc}
25–30	–45	–55	–65	–30	–40	–50	0.15	0.15	0.10	0.10
31–35	–50	–60	–70	–15	–25	–35	0.15	0.15	0.10	0.15
36–40							0.20	0.20	0.10	0.15
41–50	–55	–65	–75	–5	–15	–25	0.20	0.20	0.10	0.15
51–60	–60	–70	–80	0	–10	–20	0.20	0.25	0.15	0.20
61–70				0	–5	–10	0.20	0.25	0.15	0.20

The fundamental difference of Arc-steels from F steels is that their performance characteristics must be guaranteed by the production process. In view of the current tasks of the year-round exploration of the Northern Sea Route, development of offshore fields and coastal infrastructure, as well as transportation facilities in the Arctic regions, development of new cold-resistant Arc-steels is important and well-timed work.

As known, performance characteristics can be improved by means of efficient grain refinement [2], ensuring high metallurgical quality of steel [3] and forming a favorable deformation texture [4].

Searching for dependencies between the steel performance characteristics and structure parameters [5, 6] is a promising area of research, with the analysis of the collected data necessitating further review and improvement of the criteria for evaluation of steel performance [7, 8].

The challenging task of development of new generation low-carbon low-alloyed Arc-steels with high performance may be solved due to the balanced content of costly alloying additions in steel and special precision thermo-mechanical processing [9–13]. This became possible with the development of new techniques for recognition and quantitative analysis of structural constituents [14–16].

The purpose of this work is development of an integrated method for creation of new cold-resistant Arc-steels with the guaranteed yield strength of 355 and 390 MPa. This method is to include:

- Improvement of the chemical composition;
- Finding new scientific and technical solutions for selection of thermo-deformation rolling patterns for roughing and finishing rolling.

2. Materials and experimental technique

The material selected for the examination is low-carbon low-alloyed cold-resistant rolled steel plate 40–60 mm thick with the guaranteed yield strength of 355–390 MPa.

Mechanical tests were conducted in accordance with GOST R 52927-2015. Ductile-to-brittle transition temperature T_{cb} was determined in accordance with "Rules ..." of the Russian Maritime Register of Shipbuilding, nil-ductility temperature NDT was determined in accordance with "Rules ..." of the Russian Maritime Register of Shipbuilding and ASTM E208. Crack resistance test (CTOD test) was conducted according to BS 7448 P.1 and "Rules ..." of the Russian Maritime Register of Shipbuilding.

Examination of the structure was done by means of optical metallography on the "AxioObserverA1M" microscope and by means of automated analysis of electron back-scattering diffraction patterns (EBSD-analysis) with the use of the "Quanta 3D FEG" SEM [14]. The structure anisotropy coefficient was determined with the use of the Thixomet image analyzer in collaboration with FGAOU VO SPbPU (St. Petersburg Polytechnic University) [15]. Examination of the fine structure was done on the Tecnai G2 30 S-TWIN transmission electron microscope at the accelerating voltage of 120 kV.

3. Experimental part

3.1. Improvement of the chemical composition

Chemical composition was improved due to adjustment of the content of carbon and main alloying elements which strengthen the metal component of the interatomic bond (nickel, manganese, copper). The study included quantitative analysis of the changes in strength and impact energy at low temperatures as the content of the stated elements was varied within the range required under GOST 52927-2015 (Table 2).

Statistical analysis of mechanical properties was conducted for 70–120 steel plates of each chemical composition. The plates 40–60 mm in thickness of cold-resistance grades E and F with the yield strength of no less than 355–390 MPa were produced in industrial conditions. The most stable combination of steel strength and ductility was achieved in steel plates having a carbon content of 0.05–0.07%, a manganese content of 1.2%, a nickel content of 0.7%, a copper content of 0.2%, microalloyed with niobium.

Table 2. Influence of carbon, manganese, nickel and copper content on the stability of the strength and ductility combination in low-carbon low-alloyed steel plates.

Carbon content, %	Contents of main alloying elements	Chemistry No.	Microalloying	Mechanical properties, Min – max / average: Yield strength $\sigma_{0.2}$, MPa	Ultimate strength σ_u, MPa	Impact energy KV, J, at –60 °C: surface	center	% of poor impact energy results: surface	center
0.08–0.10	0.8% Mn+ 0.2% Ni	1	V+Nb	350–405 383	480–510 493	242–300* 294	166–300* 291	0	0
	1.2% Mn+ 0.2% Ni	2		400–460 422	485–560 516	20–300* 226	38–300* 207	4	8
	1.40 % Mn	3		444–506 481	564–637 601	118–300* 218	10–300* 126	0	28
	1.2% Mn+ 0.7% Ni	4		453–531 497	549–633 597	10–320 219	10–229 78	8	55
0.05–0.07	1.2% Mn+ 0.7% Ni	5		422–538 482	503–629 572	10–360 261	10–398 201	2	13
	1.2% Mn+ 0.7% Ni+ 0.2% Cu	6	Nb	435–530 479	550–620 576	50–296* 222	14–291 199	0	4
Requirements of GOST R 52927-2015 to steel of maximum strength grade F40W[arc]				No less than 390	510-660	No less than 50		–	

Note: * Impact energy KV at a testing temperature of – 40 °C.

3.2. Selection of the desired structural condition

Experiments have proved that in order to achieve guaranteed performance (including cold-resistance) of low-carbon low-alloyed steel plates with the yield strength no less than 355 and 390 MPa to meet the requirements to Arc-steels, it is necessary to ensure formation of fine-grained quasi-isotropic ferrite-bainite structure with similar in morphology structural constituents with the anisotropy coefficient K_a [15] no more than 0.65 and a share of lath bainite no more than 20% (Figure 1).

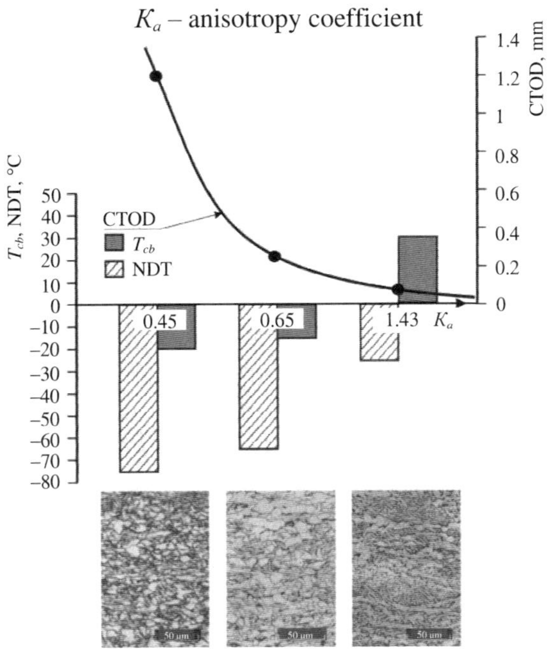

Figure 1. Influence of the morphology and anisotropy of the ferrite-bainite structure[4] of steel after thermomechanical processing on its performance characteristics.

3.3. Development of thermo-deformation rolling patterns for roughing and finishing rolling.

The work has shown that maximum austenite grain refinement due to recrystallization processes in the steels under examination, taking into account the capabilities of the modern rolling facilities, is achieved in industrial conditions by having deformation at temperatures no less than 950 °C primarily at the roughing stage – 70% of the total deformation. Decrease in deformation per pass below 10% leads to formation of individual coarse austenite grains, which, as the temperature continues to go down, transform into lath bainite. The size of the former non-recrystallized austenite grains remains unchanged, and ferrite precipitation occurs at the boundaries of such austenite grains. As a result, we obtain anisotropic structure with reduced cold-resistance.

Grain refinement at the finishing stage is achieved due to formation of well-developed subgrain austenite structure during isothermal rolling at a temperature below the recrystallization temperature (with the spread in surface temperatures between the passes not exceeding 20°C) and at 10–15% reduction per pass. It has been found that deformation during finish rolling with the surface temperatures in the range of the critical point A_{r3} or just above it (by 10–15°C) (which corresponds to the temperature of the end of deformation A_{r3}+(50–60)°C in the mid-thickness of the rolled plate) promotes formation of the most homogeneous structure with structural constituents similar in terms of their morphology, sizes and proportions (with the averaged crystal curvature GAM (Grain Average Misorientation) of 0.39–0.44° – Figure 2, (a)) in the section of 50 mm thick plates. With such deformation pattern we obtain a ferrite-bainite structure with 44–65% of quasi-polygonal ferrite and 35–50% of bainite, with the average size of structural constituents of 6–8 micrometers (Figure 2 (b)). These examinations were conducted with the use of EBSD-analysis and alpha constituents recognition method using the standard function GAM (Grain Average Misorientation – averaged measure for misorientations between the neighboring lattice points within the closed area. This measure reflects the degree of distortion of the crystal lattice), which proved efficient for quantitative assessment of bainite-martensite structures [14, 16]. Content of lath bainite is 13.5–16.0% (Figure 2 (c)).

[4] Examinations were conducted in collaboration with Kazakov A.A. and Kovalev P.V.

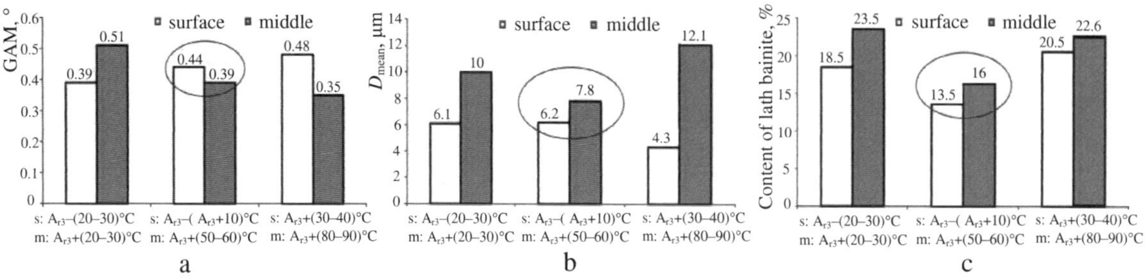

Figure 2. Influence of the temperature of the end of finish rolling: (a) on the averaged crystal curvature (GAM); (b) on the average size of structural constituents; (c) on the lath bainite content.

3.4. Industrial testing of the developed thermal and deformation rolling patterns at the roughing and finishing stages.

Industrial testing of the developed patterns during rolling of plates 50 mm thick made it possible to form quasi-isotropic structure throughout the plate thickness. The structure consisted mainly of quasi-polygonal ferrite (Figure 3 (a)) and granular bainite (Figure 3 (b)) with well-developed substructure and uniformly distributed fine carbide particles 4–12 nanometers in size (Figure 3 (c)).

Formation of this structure ensures that the required strength, visco-plastic properties and performance characteristics (T_{cb} = –15…–20°C, NDT= –65…–70 °C, CTOD^{av}=1.16 mm at testing temperature –60 °C for base metal) are achieved.

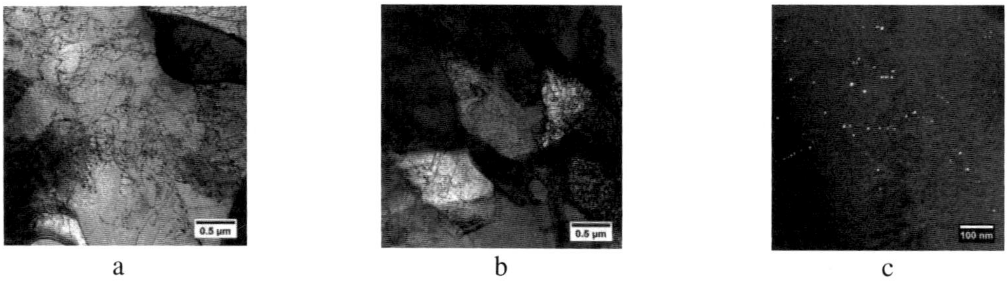

Figure 3. Structure of low-carbon low-alloyed Arc-steel with the yield strength no less than 315–390 MPa[5]: (a) quasi-polygonal ferrite; (b) granular bainite; (c) fine carbides $(120)_{Fe3C}$ in granular bainite.

4. Conclusions

1. It has been shown that the most stable combination of the required strength and cold-resistance is found in rolled steel plates containing (0.05–0.07)%C+1.2% Mn+0.7% Ni+0.2% Cu.
2. In order to ensure that steel plates meet the performance characteristics required for Arc-steels, including cold-resistance, it is necessary to form fine ferrite-bainite structure with the anisotropy coefficient K_a not exceeding 0.65 and a share of lath bainite no more than 20%.
3. We have developed thermal and deformation rolling patterns for roughing and finishing stage that ensure formation of fine quasi-isotropic ferrite-bainite structure during industrial production of Arc-steel plates with the guaranteed yield strength of 355–390 MPa.

References

[1] Gusev M A, Ilyin A V and Larionov A V 2014 *Shipbuilding* **5** 39–3
[2] Jia T, Zhou Y, Jia X and Wang Z 2017 *Metall. Mater. Trans.* A **48A** 685–96

[5] Examinations of fine structure were performed in collaboration with Mikhailov M.S.

[3] A Ghosh, A Ray, D Chakrabarti and C L Davis 2013 *Mater. Sci. Eng.* A **561** 126–35
[4] Masoumi M and Gomes de Abreu H F 2015 *Technical contribution to the 52 ABM Week (Brazil)*
[5] Thridandapani R R, Misra R D K, Mannering T, Panda D and Jansto S 2006 *Mater. Sci. Eng.* A 285–91
[6] Hu J, Du L X, Zang M, Yin S J, Wang Y G, Qi X Y, Gao X H and Misra R D K 2016 *Materials Characterization* **118** 446–53
[7] Ilyin A V, Mizetsky A V and Filin V Yu 2005 *Material science questions* **1 (41)** 55–68
[8] Danilov G I, Ilyin A V, Leonov V P and Fedorova T A 2005 *Material science questions* **1 (41)** 69–78
[9] T Gladman 1997 *The physical metallurgy of microalloyed steels.* Book 615 (London: The Institute of Materials)
[10] Kruglova A A, Orlov V V, Sych O V and Khlusova E I 2013 *Metallurgist* vol 57 **1-2** 113–122
[11] DeArdo A J 2001 *Ironmaking and Steelmaking* vol 28 **2** 138–44
[12] Orlov V V, Malyshevsky V A, Khlusova E I and Golosienko S A 2014 *Steel* **9** 79–88
[13] Goli-Oglu E A and Bochkarev Yu A 2014 *Steel* **9** 71–8
[14] Zisman A A, Petrov S N and Ptashnik A V 2014 *Metallurgist* **11** 91–5
[15] Kazakov A A, Kazakova E I, Kiselev D V and Motovilina G D 2009 *Ferrous metals* **12** 12–5
[16] Kang J U, Kim D H, Baik S-I, Ahn T-H, Kim Y-W, Han H H, Oh K H, Lee H-C and Han S H 2011 *ISIJ International* vol 51 **1** 130–6

Acknowledgements. The work was carried out under the state contract with the Ministry of Industry and Trade of the Russian Federation No. 16411.1810190019.09.003 dd. 20th October, 2007.

Numerical Simulation of Roller Levelling using SIMULIA Abaqus

K A Trusov[1], P A Mishnev[1], O V Kopaev[1] and D V Nushtaev[2]

[1] JSC «Severstal Management», Mira street 30, Cherepovets, RU-162600, Russia
[2] TESIS Ltd., Unnatov street 18, Moscow, RU-125083, Russia

E-mail: ka.trusov@severstal.com

Abstract. The finite element (FE) 2D-model of roller levelling process is developed in the SIMILIA Abaqus. The objective of this paper is development FE-model and investigation of adjustable parameters of roller leveller together with elastic-plastic material behaviour. Properties of the material were determined experimentally. After levelling, the strip had a residual stress distribution. The longbow after cutting is predicted too. Recommendation for practical use were proposed.

1. Introduction

Numerical prediction of residual stresses become of practical importance when the laser cutting of strips is applied [1]. High values of residual stresses lead to deformation of strips (longbow) during and after laser cutting.

There are many operations in metallurgy before cutting. It is slab heating, slab rolling until given thickness, cooling, coiling, uncoiling, levelling and strip cutting. After each operation, the material accumulates an internal (residual) stress that is distributed throughout the strip [2]. When laser cutting is applied, the balanced stress state of the material is disturbed and it leads to the longbow defect. Sheet metal often shows shape defects, which is not complying with the increasing requirements for the quality of products needed to satisfy high demands of customers [3].

In general, there are two ways to flatten sheet:

1) In roller levelling, the strip is bent through a series of rolls, moving from heavy penetration at the entry and at the end distance between the top and bottom contact rolls equals the strip thickness.

2) In tension levelling, large-diameter bridle rolls at the material exit rotate slightly faster than the work rolls behind it, pulling the strip from the source coil under very high tension.

The setting up of such machines is extremely complex and mainly depends on the operator's experience.

Some research has been done to address the longbow problem. A few works [4, 5] describe general models for roller levelers but these models could not provide precision settings for each machine. For example, the authors of work [3] developed a precision mathematical model for setting up a roller leveler that allows making flat sheet in on-line mode.

A recent increase in computing power enabled simulation to be used for production processes to the best advantage. The problem of residual stress calculation could be solved by using FE analysis. Influence of residual stresses on shape defects of strips is examined in articles [6-11]. But the authors don't analyze shape of the sheet having been cut into narrow blanks.

IOP Conf. Series: Materials Science and Engineering **287** (2017) 012014 doi:10.1088/1757-899X/287/1/012014

Works [12-14] describe simulation of laser cutting and prediction of thermal stresses in the cutting area.

This paper gives details on FE 2D-simulation of levelling in SIMILIA Abaqus and analysis of longbow defect after cutting.

2. FE-simulation description

For this analysis, leveler with 15 rolls, 8 over 7, is used. The simulation is carried out for 15 rolls (8 top, 7 bottom) leveler (Figure 1). Input data for simulation is:

- strip (width, thickness, material hardening law);
- position of work rolls.

The simulation is carried out for an incoming strip with a radius of curvature equal to the coil lap (500mm). The strip has 10 elements along thickness. Element type used for the strip is 4-node bilinear plain strain with reduced integration (CPE4R). The strip is assumed to be in a stress free state before uncoiling.

Rolls are simulated by discrete rigid elements.

The tangential contact between rolls and steel strip were simulated with a coefficient of static friction μ_{stat}=0.25 and a coefficient of kinetic friction μ_{kin}=0.12.

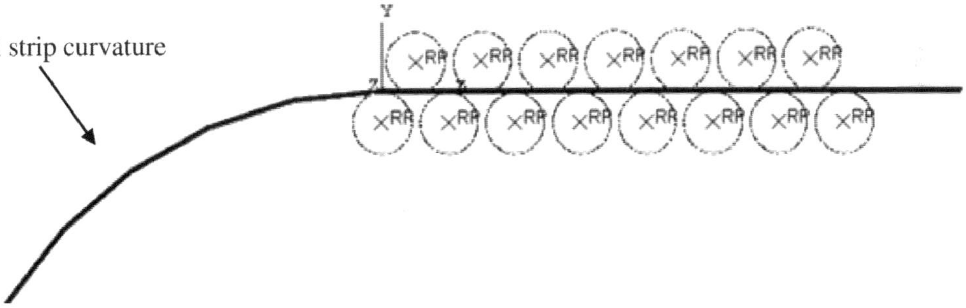

Figure 1. Assembly initial state.

The material properties used for simulation are given in Table 1 and Figure 2. These properties were determined experimentally for steel grade S355MC, 3mm thickness.

Table 1. Material properties.

Steel grade	Elastic modulus, GPa	Yield Strength, MPa	Tensile Strength, MPa
S355MC	206,8	380	450

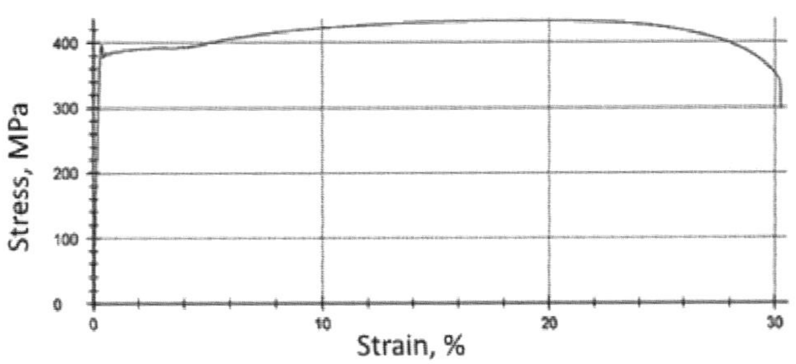

Figure 2. Tensile test curve for steel grade S355MC (3mm).

IOP Conf. Series: Materials Science and Engineering **287** (2017) 012014 doi:10.1088/1757-899X/287/1/012014

The analysis includes 6 steps:

- Step 1 (Dynamic, Implicit): Upper rolls move vertically and penetrate the strip. Bottom rolls remain in their position (Figures 3 and 4).

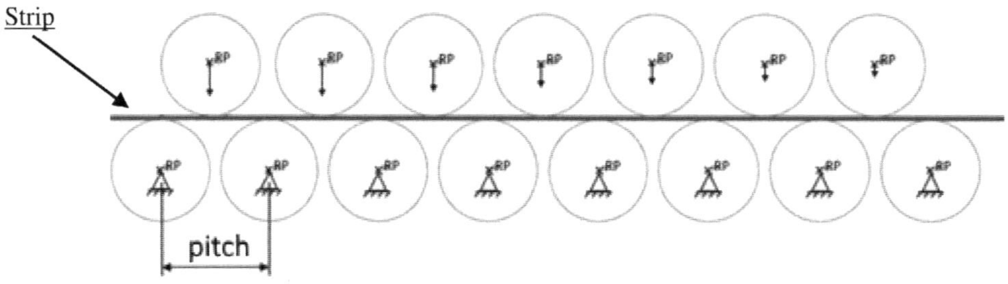

Figures 3. Initial State at Step 1.

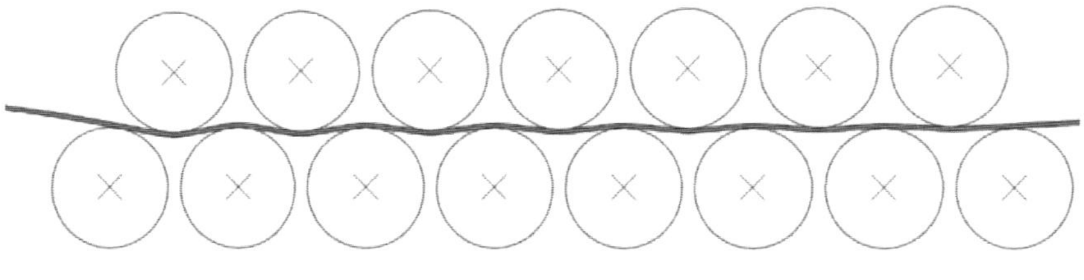

Figure 4. Upper rolls are brought into working position.

- Step 2 (Static, General) is needed for reset velocity and acceleration vectors.
- Step 3 (Dynamic, Implicit). Strip levelling
- At start of the step to the rolls applied velocity and after 1 second, rolls achieved working speed.
- Step 4 (Static, General): is needed for reset velocity and acceleration vectors;
- Step 5 (Dynamic, Implicit): Cutting.

"Model change" function is used and part of elements are deactivated. After that, bend of the strip under the influence of residual stresses is analyzed.

3. Results and discussion

Residual stress distributions in the strip are modeled and longbow defect after cutting is simulated. Various settings of roller leveler are analyzed. This analysis is presented in Figure 5 and Table 2.

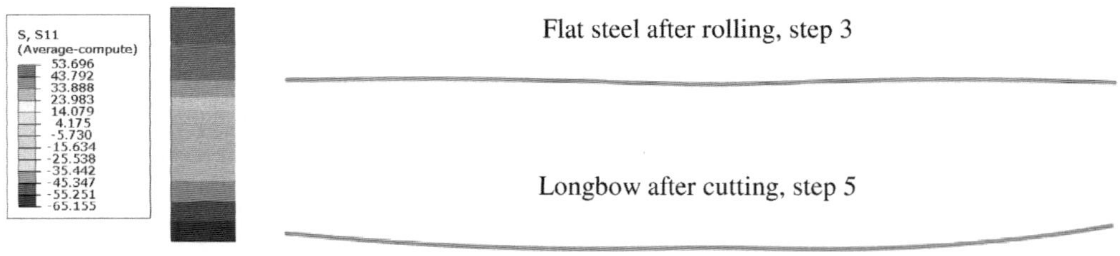

Figure 5. Longitudinal residual stress distribution through thickness before Step5 and longbow simulation after Step 5.

Table 2. Variants of various setting ups of work rolls.

№	Upper roll penetration of strip		Gap between rolls	Stress difference between surfaces of sheet, MPa	Schematic longbow defect
	First roll, mm	Last roll, mm			
1	3	3	0	374	
2	1.5	0	1.5	118	
3	3	1.5	1.5	304	
4	4	2.5	1.5	311	
5	5	3.5	1.5	344	
6	8	5	3	401	
7	5	0.5	4.5	151	
8	6.5	2	4.5	303	
9	6.5	0.3	6.2	113	

Results of computation and analysis:

1) Penetration of first upper roll shall have a maximum value (leveler with maximum gap value 4,5 mm is modelled) to provide plastification rate from 60 to 80%.

2) Penetration of last upper roll shall be as small as possible and equal to the thickness of the strip.

These recommendations were proposed for validation in experimental levelling tests in PAO «Severstal».

The main objective of this work is to build a numerical simulation model of steel sheet levelling and analyze a longbow defect after cutting. The model allows for a virtual determination of leveler settings based on adjustable bending of the levelling rolls.

References

[1] Milenin A, Kustra P, Kuziak R and Pietrzyk M 2014 *Procedia Engineering* **81** 108–13
[2] Totten G E (Ed) 2002 *Handbook of residual stress and deformation of steel* (USA: ASM international)
[3] Behrens B A, El Nadi T and Krimm R 2011 *Journal of Materials Processing Technology* **211** **(6)** 1060–8
[4] Nedorezov I V 2003 Mathematical simulation of the levelling processes on multiroller machines (Ekaterinburg: AKVA-press)
[5] Shinkin V N 2015 *Young Scientist* **10** 361-6
[6] Morris J W, Hardy S J, Lees A W and Thomas J T 2001 *Ironmaking steelmaking* **28 (1)** 44–52
[7] Grüber M, Oligschläger M and Hirt G 2015 *Key Engineering Materials* vol 651–653 pp 1023–1028
[8] Kaiser R Hatzenbichler T, Buchmayr B and Antretter T 2014 *Materials Science Forum* (Switzerland: Trans Tech Publications) vol 768 pp 456–63
[9] Zhou C L, Jiang Z Y, Wang G D, Xie D G and Zhang Y A 2010 *Advanced Materials Research* (Switzerland: Trans Tech Publications) vol 97 pp 71–4
[10] Tadić N and Mišović M 2007 *Metalurgija* **13 (4)** 251–57
[11] Marnette J, Rolfe B, Hodgson P and Weiss M 2013 *AIP Conf. Proc.* vol 1567 **1** pp 452–55
[12] Arif A F M and Yilbas B S 2008 *The International Journal of Advanced Manufacturing Technology* **37 (7-8)** 698-704
[13] Arif A F M, Yilbas B S and Aleem B A 2009 *Laser Technology* **41 (3)** 224–32
[14] Dyachenko U V, Makovetskii A V and Hitrih E E 2014 *Open information and computer integrated technologies* **63** 104–13

IOP Conf. Series: Materials Science and Engineering **287** (2017) 012015 doi:10.1088/1757-899X/287/1/012015

Optimization of Thermal and Deformation Effect during Plastic Deformation and Thermal Treatment of Hot Rolled Heavy Wall Pipes Made of Medium Carbon Martensitic Steel

A I Ziza[1], V V Tsukanov[1]

[1] Kurchatov Institute Research Centre – Prometey Central Research Institute of Structural Materials, Shpalernaya Street 49, Saint Petersburg, RU-191015, Russia

E-mail: oknir@crism.ru

Abstract. The method of stress relaxation in isothermal aging after deformation is employed to determine recrystallization temperature range (T_{rec}) for 38HN3MFA (35HN3MFA) steel grade. The paper shows that billet final deformation at the temperature lower than T_{rec} followed by annealing including heating and aging at T_{rec} enables austenite grain refinement.

1. Introduction

Improved ductility, impact strength and failure resistance for a specified level of material strength is obtained by increasing degree of structural fineness [1]. Size of transformed structure elements is directly related to initial austenite grain. According to [2] martensite can form three packages within prior-austenite grain boundaries, the packages further divide into blocks and sub-blocks. Before $\gamma \rightarrow \alpha$ transformation starts recrystallization of deformed γ phase is used for austenite grain refinement [3]. However, martensitic steels show a tendency to structural heredity representing a phenomenon when martensite blocks are formed within prior austenite grain, which is recovered when being reheating [4]. Structural heredity is prevented by using recrystallization annealing [5], which includes heating up to austenite recrystallization temperature and aging at this temperature.

In this case determination of precise recrystallization temperature range plays a key role for structure refinement. State-of-the-art testing equipment allows determining T_{rec} by using stress relaxation method [6].

Besides, when material deformation is performed at the temperature lower than T_{rec}, austenite structure can split into disordered fragments [7]. Boundaries of such fragments as well as grain boundaries are locations of α phase primary nucleation [8]. Besides, martensite can partially inherit austenite disordered structure.

Hot rolled heavy wall pipes with a wall thickness of 65 mm (before machining) are widely used for manufacturing high pressure cylinders with a working pressure of 39.2 MPa. These pipes are made of high strength medium carbon martensitic steel of 38HN3MFA and 35HN3MFA grades, their chemical composition is shown in Table 1.

Table 1. Chemical composition of 38HN3MFA and 35HN3MFA grades.

Steel grade	C	Mn	Si	Cr	Ni	Mo	V	P	S
38HN3MFA	0.33 0.40	0.25– 0.50	0.17– 0.37	1.20– 1.50	3.00– 3.50	0.35– 0.45	0.10– 0.18	≤ 0.025	≤ 0.025
35HN3MFA	0.34– 0.39	0.20– 0.40	0.17– 0.37	1.10– 1.40	2.75– 3.25	0.30– 0.40	0.08– 0.15	≤ 0.025	≤ 0.020

Pipe billet for a high pressure cylinder is made by piercing an ingot and its further consistent drawing over the mandrel gradually reducing the diameter of rings and mandrel. After the billet undergoes preliminary thermal and mechanical treatment, necks are forged. After final thermal treatment (quenching + high tempering) the structure of cylinder steel is tempered martensite with blocks size corresponding to numbers 7–8 according to GOST 5639. Mechanical properties of steel for cylinders shall meet the following requirements: tensile strength $\sigma_{ts} \geq 1128$ MPa, yield stress $\sigma_{0.2} \geq 981$ MPa, elongation $\delta_5 \geq 11$ %, impact strength $KCU^{+20} \geq 68{,}7$ j/cm^2.

History of manufacturing and using cylinders [9] shows that cylinder material has a significant variation of impact strength values and strength properties, which cannot ensure trouble-free use. There were also some cases when impact strength was lower than acceptable minimum level while strength properties significantly exceeded standard requirements.

Another concern is that the mechanical properties of cylinder material are tested using witness samples, which are pieces of billets thermally treated together with cylinders in the same batch. Moreover, the difference between mechanical properties of cylinder material and witness sample material can reach 15% [9].

Variations in mechanical properties of material and functional reliability of cylinders can be improved by refining cylinder fabrication practices and particularly billet production process. This is since the best mechanical properties can be obtained by controlling structure of the material starting from billet deformation followed by preliminary thermal treatment.

Thus, the purpose of this work is optimization of thermal and deformation parameters during deformation and preliminary thermal treatment of pipe billets to obtain refined structure and best mechanical properties for trouble-free use of high pressure cylinders.

2. Selection of preliminary thermal treatment

38HN3MFA and 35HN3MFA grades tend to structural heredity, which makes it difficult to obtain highly refined homogeneous structure. Structural heredity results in the fact that coarse grain with a martensite or bainite ordered structure will not become finer when heated anew higher than Ac_3 point, but will replicate the initial one in terms of size and orientation [10]. Impact strength is a structure-responsive feature and heavily depends on the size of structural constituents. This is because failure occurs according to "weak link" principle, fracture growth along fine grains boundaries requires continuous change of its direction, therefore fracture growth along coarse grains boundaries is more energy-efficient. Structural heredity can manifest in fracture growing along the boundaries of prior austenite grains, despite the actual refined grain [10].

Previous works [5] show that recrystallization annealing is practical for grain refinement of Cr-3Ni-Mo-V steels with high austenite resistance. Recrystallization annealing includes heating followed by aging at the recrystallization temperature (T_{rec}). This thermal treatment contributes to emerging of new finer austenite grains. In this case preliminary deformation and phase hardening play a crucial role for the process, as they are main driving forces for recrystallization. Therefore, material recrystallization temperature shall be defined for grain refinement.

The second reason for defining (T_{rec}) is related to control of deformation finishing temperature. During plastic deformation at a temperature lower than T_{rec} austenite hardening occurs and substructure appears, that is deformation at a temperature lower than T_{rec} contributes to formation of austenite subgrain structure caused by deformation. It is inherited by transformed austenite structure, which causes formation of optimal structure – lath martensite. At the same time last pass during billet deformation at a temperature higher than T_{rec} can cause partial grain growth due to secondary recrystallization running at too high temperature of billet heating. In this case the driving force of secondary recrystallization is surface energy of grains, as metal tends to more balanced state, which goes along with reduction of total length of grains boundaries, that is grain coarsening.

3. Recrystallization temperature determination

Gleeble 3800 thermomechanical simulator was used to determine (T_{rec}). Cylindrical samples of Ø10×15 mm were heated up to 1220 °C and cooled down to the supposed T_{rec} (860, 880, 940, 960 °C) – deformation temperature, at which uniaxial compression followed by ageing at this temperature was performed. The method of stress relaxation [11] was employed to observe austenite recrystallization kinetics.

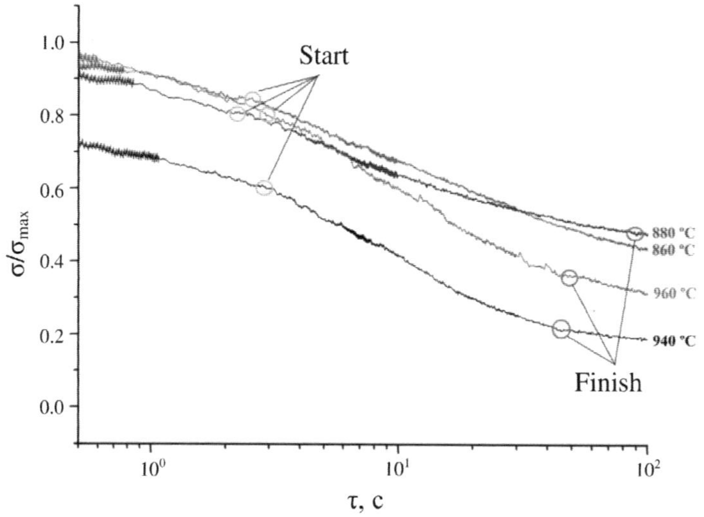

Figure 1. Diagrams of stress relaxation after deformation completion at temperatures of 860, 880, 940, 960 °C.

Figure 1 shows stress relaxation diagrams for different temperatures demonstrating dependence of samples softening on aging, typical for running of the recrystallization process. Apparently, recrystallization doesn't stop at a temperature of 860 °C during time interval selected. Temperature of 860 °C can be considered as a lower level of recrystallization temperature range. Recrystallization stops as quick as possible during isothermal aging at temperatures of 940 and 960 °C. Therefore 940 ± 20 °C range can be considered as the desired T_{rec} range.

4. Microstructure examination

Gleeble 3800 thermomechanical simulator was used to simulate deformation patterns with a temperature of last pass lower than (T_{rec}) range at 880 °C and higher than (T_{rec}) range at 980 °C and subsequent preliminary (annealing at 940 °C and high tempering at 690 °C) and final (quenching at 880 °C and double tempering at 610 °C) thermal treatment. Samples were etched under vacuum according to the procedure [12] to find out austenite grain size and assess efficiency of measures taken to have finer structure blocks. Figure 2 shows samples microstructure. When deformation temperature of the final pass exceeds (T_{rec}) range, an average austenite grain size corresponds to number 9 according to GOST 5639 (Figure 2 (a)).

An average austenite grain size of the sample deformed with a final pass temperature exceeding (T_{rec}) range corresponds to number 9 according to GOST 5639 (Figure 2 (a)). When deformation temperature of the final pass is lower than (T_{rec}) range, an average austenite grain size corresponds to number 10 according to GOST 5639 (Figure 2 (b)).

However, it should be noted that, microstructures were compared after simulation of full heat treatment cycle to obtain representative values against earlier test results of high pressure cylinder material.

a

b

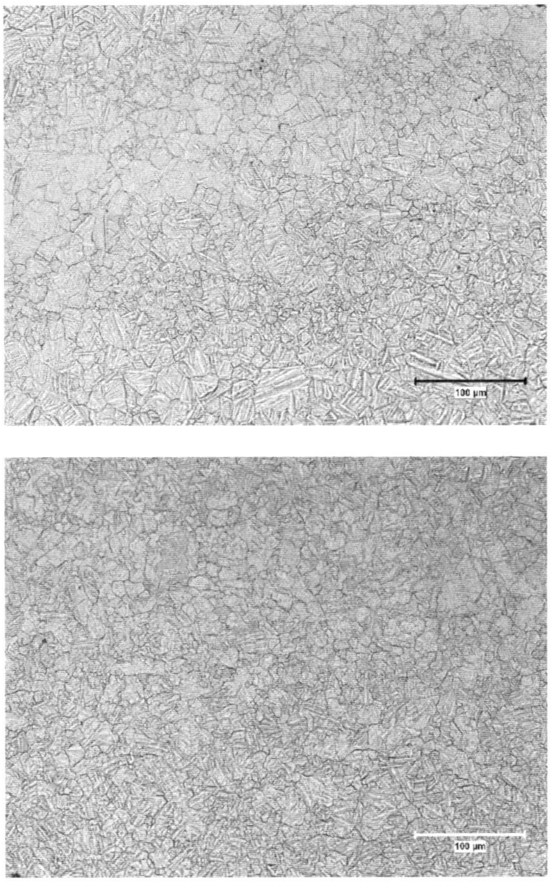

Figure 2 Sample material microstructure after deformation and thermal treatment:
(a) final deformation at 880 °C;
(b) final deformation at 980 °C, × 200.

For further assessment of grain refinement efficiency, the samples were etched under vacuum after having been cut out of high pressure cylinders fabricated both under old and improved technology, according to the following above-mentioned recommendations: Final stage of billet deformation at a temperature lower than (T_{rec}) range followed by annealing at a temperature within (T_{rec}) range. Figure 3 shows samples microstructure. Austenite grain size of material for cylinders made under old technology corresponds to numbers 7–8 according to GOST 5639 (Figure 3 (a)). In this case maximum grain size will reach 88.3 µm. Austenite grain size of material for cylinders made under improved technology corresponds to numbers 9–10 according to GOST 5639 (Figure 3 (b)). In this way, the structure can be considered more homogeneous with a maximum grain size of 43.7 µm.

a

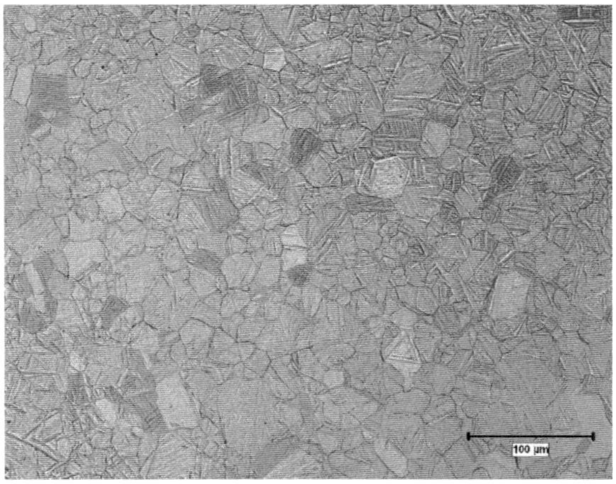

b

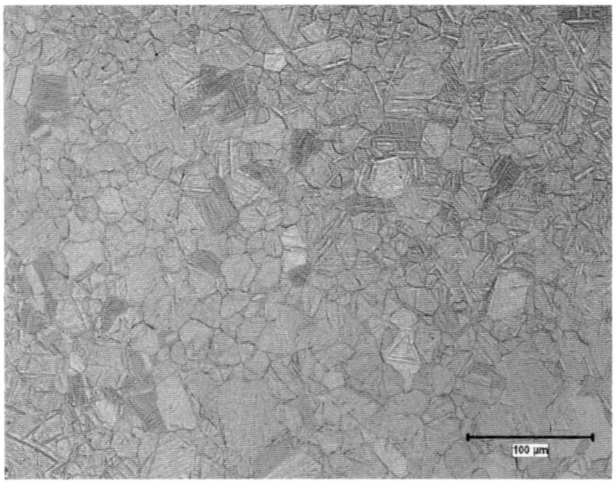

Figure 3 Microstructure of high pressure cylinder material after final thermal treatment:
(a) old technology
(b) improved technology, × 200.

4. Conclusions

1. State-of-the-art research simulators provide manufacturing process modeling, which enables to obtain metal structure like this obtained in production environment.

2. Final deformation at the temperature lower than recrystallization temperature followed by annealing including heating and aging within recrystallization temperature range contributes to refinement of structural elements for 38HN3MFA and 35HN3MFA steel billets.

References

[1] Moroz L S 1984 *Mechanics and Physics of Material Deformation and Failure* (Leningrad: Mashinostroyeniye) p 225
[2] Morito S, Saito H, Ogawa T, Furuhama T and Maki T 2005 *ISIJ International* vol 45 **1** 91–94
[3] Miao C L, Shang C J, Zhang G D and Subramanian S V 2010 *Mater. Sci. Eng.* A **527**, 4985
[4] Nakada N, Tsuchiyama T, Takaki S and Hashizume S 2007 *ISIJ International* vol **47 10** 1527–32
[5] Tsukanov V V and Ziza A I 2015 *Issues of Materials Science* **2**(82) 9–16
[6] Perttula J S and Karjalainen L P 1998 *Materials Science Technologies* **14** 626-30
[7] Kodzhaspirov G Ye, Rybin V V and Apostolopoulos H 2007*Metal Science and Heat Treatment* **1** (619) 30–34
[8] Quispe A, Medina S F, Gomez M and Chaves J I 2007 *Materials Science and Engineering* A **447** 11–18

IOP Conf. Series: Materials Science and Engineering **287** (2017) 012015 doi:10.1088/1757-899X/287/1/012015

[9] Zakharenko Yu V, Ziza A I, Ilyin A V and Tsukanov V V 2017 *Metalworking* **4** (100) 29–36
[10] Schastlivtsev V M, Zeldovich V I, Mirzayev D A and others 2008 *Development of V.D. Sadovsky ideas. Collected works* ed M A Filippov and Yu V Kaletina (Yekaterinburg: Institute of Metals Physics, Ural Branch of Russian Academy of Science) p 409
[11] Zisman A A, Soshina T V and Khlusova Ye I 2012 *Notes on Materials* vol 2 **1** 3–8
[12] Soshina T V, Zisman A A and Khlusova Ye I 2013 *Metallurgist* **2** 63–70